Armand de Watteville

A Practical Introduction to Medical Electricity

Armand de Watteville

A Practical Introduction to Medical Electricity

ISBN/EAN: 9783744791021

Printed in Europe, USA, Canada, Australia, Japan

Cover: Foto ©berggeist007 / pixelio.de

More available books at **www.hansebooks.com**

A PRACTICAL INTRODUCTION

TO

MEDICAL ELECTRICITY

A PRACTICAL INTRODUCTION

TO

MEDICAL ELECTRICITY

BY

A. DE WATTEVILLE, M.A., M.D., B.Sc.

MEMBER OF THE ROYAL COLLEGE OF SURGEONS OF ENGLAND.
PHYSICIAN IN CHARGE OF THE ELECTRICAL DEPARTMENT AT ST. MARY'S HOSPITAL, LONDON.
LATE ASSISTANT PHYSICIAN TO THE HOSPITAL FOR EPILEPSY AND PARALYSIS; AND
ELECTROTHERAPEUTICAL ASSISTANT TO UNIVERSITY COLLEGE HOSPITAL.

SECOND EDITION

WITH EIGHTY-NINE ILLUSTRATIONS AND PLATES OF THE MOTOR POINTS

LONDON
H. K. LEWIS, 136 GOWER STREET, W.C.
1884

PRINTED BY
H. K. LEWIS, 136 GOWER STREET,
LONDON, W.C.

PREFACE.

This is a new book rather than the second edition of an old one; I have found it necessary not only to add much new material, but to rewrite the greater part of what I allowed to remain. The list of instruments which in the first edition gave the impression of an illustrated catalogue, and occupied much space uselessly, has been considerably reduced so as to keep the book within reasonable limits.

I still insist upon *measurements of current strength* as the essential condition of a rational application of electricity to medicine. It is gratifying to me that the proposition I first made in 1878 of adopting the milliweber, now called milliampere, as the electro-therapeutical unit, has received the sanction of the special committee of the International Congress of Electricians. Absolute galvanometers are becoming very generally used on the Continent, especially in France and Germany.

Of equal importance with an accurate knowledge of the actual electrical quantities we are using, is a clear understanding of the *distribution* of those quantities among the tissues we wish to influence.

This electrical distribution, or diffusion, depends largely upon the extent of the surface through which the current is introduced into the body; I therefore propose the adoption of Prof. Erb's set of *standard electrodes* so as to fix once for all the meaning of the terms "small," "large," etc., hitherto applied to them in a loose manner. The second factor of current diffusion being the relative *position* of the electrodes on the body, I have laid more emphasis upon this topic, supplementing my previous diagrams with the excellent figures borrowed from the author just mentioned.

The chapter on electrophysiology will be found to be new both as to matter and form. It embodies the results of observations made on man, which are presented in the shape of experiments or exercises, the repetition of which the student will find the shortest way of acquiring a complete mastery over the manipulations necessary for the practice of diagnosis and treatment by electricity.

The subject of electrodiagnosis has undergone no marked development of late. The description of the phenomena of morbid reactions has been merely expanded; but I have insisted much more fully upon the difficulties of electrophysiological experiments on man, for such is the performance of an electrodiagnostic investigation, so that this part of chapter III. may be taken as a continuation of the preceding one.

With reference to electrotherapeutics proper, the reader will observe that the whole chapter on treatment has been written by myself. The special part of it was in the former edition a translation from the French of the corresponding portion of my friend Dr. Onimus' *Guide Pratique*. My views and methods have undergone considerable changes during the past years, and the opportunities I have had of bringing them to the test of practice have been numerous enough to justify me, I hope, in expressing myself with some confidence on this part of the subject.

I trust that in speaking of the curative value of electricity I have been able to keep a happy medium between excessive enthusiasm and premature discouragement. We have still much to learn and to improve with reference to the modes of electrisation in disease. I have endeavoured to give the rationale of the methods, as well as directions for carrying out the several applications of the currents to disease, so that the physician should be able to modify his treatment and adapt it to the peculiarities of the cases which come before him, a task by no means so simple as some appear to think.

Finally I have, after some hesitation, added a paragraph on the electrolysis of tumours intended to rationalise the method and stimulate further researches on the subject, rather than to endorse, and give a complete account of, the results obtained—or at least stated to have been obtained—in this field of surgical work.

Many new illustrations will be found in the second edition, among which several instructive diagrams (figs. 1—7, 22—25) borrowed from M. Niaudet's *Traité de la Pile*, of which a translation has, I believe, been published by Messrs. Spon. My friend, Prof. Erb, has kindly allowed me to use some diagrams of the Reactions of Degeneration (figs. 79—83), and of the current-diffusion in the human body (figs. 84—88) taken from his great work on *Elektrotherapie* (Volume III. of Ziemssen's series of General Therapeutics) of which I am preparing an English edition for Messrs. Smith, Elder and Co.

I have abstained from giving any bibliographical references, partly for the sake of economising space, partly because the forthcoming treatise by Prof. Erb will contain a very full list of them. For similar reasons, I have not introduced any illustrative cases into the chapter on treatment; my object is to enable the reader to observe and act for himself, rather than to astonish him by narrations of any of the wonderful cures—self administered testimonials—so dear to the specialist. Electricity has now won a recognised place in medicine in spite of the dangerous praises of enthusiasts; but whilst admitting that its therapeutical value must be determined by the empirical results of clinical experience only, I have endeavoured to remain faithful to the strictly scientific modes of thought required by such a physical agent in its study and applications.

One word in conclusion. There are several considerable practical difficulties which will no doubt, at least for some time to come, prevent electricity from becoming really popular with the medical profession. Foremost stands the absence of any theoretical and practical teaching in our schools: it is the object of this little book to supplement this want in some degree. Next comes the question of the apparatus required, which kind is the best, where to obtain it, and how to keep it in working order. We in England are at a decided disadvantage with reference to our supply of electrical instruments. Abroad there is no dearth of professed electricians prepared to supply the requirements of the physician; here the latter has hitherto had to depend upon the surgical instrument maker from whom it is difficult to obtain anything beyond the adopted battery of his predilection. I have felt this want the more acutely that I have been constantly obliged to answer inquiries on this subject, from medical men in trouble.

I am glad therefore to mention here, in addition to Mr. Schoth's name, that of Mr. Thistleton of 1 Old Quebec Street, Oxford Street, who has quite recently announced his intention to devote himself solely to the manufacture of medico-electrical apparatus.

A DE WATTEVILLE.

30 Welbeck Street, W.
January, 1884.

CONTENTS.

Chapter I.

ELECTROPHYSICS.

	Page
Electrification	1
Potential	2
Current	3
Electromotive Force	5
Resistance	10
Electrical Units	12
Graphical Method	16
Measurement of Electromotive Force	18
Measurement of Resistance; Rheostats	20
Measurement of Current Strength	22
Arrangement of Elements	32
Constancy of Currents	38
Derived Currents	40
Density of Current	41
Diffusion of Current—Distributions of Potentials	42
The Human Body as a Conductor	43
Mechanical Effects of the Current	48
Physical Effects	49
Chemical Effects	49
Cells and Batteries	51
Amalgamation	55

The Peroxide of Manganese Element	56
The Chloride of Silver Element	59
The Sulphate of Copper Element	62
The Sulphate of Mercury Element	64
The Bunsen and Grove Element	65
Single Fluid Elements	66
Polarisation Cells	68
Faradism	69
Induction Apparatus	76
Accessories—Current Graduation—Rheostat and Collector	82
The Plug Collector	84
The Sledge Collector	85
The Dial Collector	87
Rheophores	88
Connections	89
Electrodes	89
Interruptors	92
Current Reverser or Commutator	94
Current Alternator, Reverser, and Combiner	95

Chapter II.

ELECTROPHYSIOLOGY.

Motor Nerves—Galvanic Excitation	97
Motor Nerves—Faradic Excitation	109
Electrotonus	110
Muscle	113
Sensory Nerves	115
Polarisation of the Human Body	119

Chapter III.

ELECTRODIAGNOSIS.

	Page
Electrodiagnosis and Prognosis of Motor Nerve and Muscle Lesions	120
The Comparison of Excitations	135
The Conditions of Equal Excitations	138
The Galvanometer in Diagnosis	140
Other Fallacies of Diagnosis	142
The Practice of Electrodiagnosis	143

Chapter IV.

ELECTROTHERAPEUTICS.

A. General—Introductory Remarks	146
Influence of Direction and of Pole	147
Choice of Current	150
Posology	151
Choice and Position of Electrodes	153
Practical Remarks on Electrisation	156
Galvanisation	158
Faradisation	160
General Electrisation	162
Galvano-Faradisation	163
Subaural Galvanisation	164
B. Special Electrotherapeutics	165
Hypokinesis:—Paralysis—Paresis	165
Hyperkinesis—Spasm	168
Anæsthesia	171

	Page
Hyperæsthesia—Neuralgia—Pain	172
Electrisation of the Brain	174
Electrisation of the Spinal Cord	178
Functional Cerebro-Spinal Diseases	180
Diseases of the Peripheral Nerves	184
Toxic Paralyses	187
Diseases of the Eye	188
Diseases of the Ear	188
Neuroses of the Heart and Lungs	190
Diseases of the Abdominal Organs	190
Diseases of the Muscles and Joints	193
Diseases of the Genito-Urinary Organs	194
Diseases of the Skin	196
Appendix—Electrolysis of Tumours	198

ERRATA.

Page 5, last line but one, *for* "KC and AC" *read* "KC and CC".
" 30, Note, *for* "HSO$_4$" *read* "H$_2$SO$_4$".
" 190, Note, *for* "2 mm" *read* "4 mm".

A PRACTICAL INTRODUCTION

TO

MEDICAL ELECTRICITY.

CHAPTER I.

ELECTRO-PHYSICS.

ELECTRIFICATION.

Under certain circumstances bodies become possessed of properties, among which is the power of inducing similar properties on neighbouring bodies, and of attracting or repelling the latter. We say that a body possessing these properties, is *electrified*.

The electrification of a body may be more or less considerable; and by a process of mental abstraction, we attribute this electrified condition to the presence in it of a greater or less quantity of an incompressible fluid which we call electricity.

In this way, electrification becomes susceptible of calculation, changes in its intensity being explained by the addition or subtraction of certain quantities of fluid called electricity.

The usual methods of electrifying a body are:—1. Friction; 2. Contact; 3. Induction.

According to the mode of its production, electricity displays some of its properties in a more marked degree than the others.

In frictional electricity (franklinism), those properties which are designated under the name of *static*, are most highly developed. As the word implies, electro-statics deals with the phenomena of electricity at rest. Owing to the very limited applicability of static electricity to medicine, it will not be studied in these pages, and the reader is referred for further information to any of the usual text-books on electricity.

Electricity in movement, or current-electricity, is endowed with but faintly-marked static properties. Its phenomena are chiefly of the *dynamic* order; and concern us more particularly. Electro-kinetics (κινέω, to move) embraces the study of *galvanism* (also called voltaism, contact-electricity, the chemical or the constant current), and *faradism* (also called the induced, the magneto-electric or the interrupted current).

The galvanic current is generated by the contact of dissimilar substances, in presence of chemical action (or of heat: thermo-electricity).

The faradic current is generated by the inductive influence of galvanism and magnetism, in presence of variations in the intensity of this influence. Though the fundamental laws of electro-kinetics apply equally to galvanism and to faradism, yet their modes of production and external manifestations are sufficiently different to require that we should describe them separately. We shall combine our account of galvanism with the discussion of the laws governing electric currents, because the full verification of the latter through measurement is possible only by this form of electricity. We shall then touch upon faradism, showing the main peculiarities which differentiate it from galvanism.

I. GALVANISM.—POTENTIAL.

We say there is a difference of Potential between two places whenever electricity tends to move from one of these places to the other.

Thus, there is a close parallelism between electricity and water, or any other material fluid. If we wish to produce a flow of water from A to B, we must make A higher than B. In precisely the same manner, if we wish to have a flow of electricity between two points, we must create a difference of level between then. The point from which electricity (or water) flows, must be at a higher potential (or level), than the point to which it flows.

When we speak of the potential of a point we mean the difference between the potential, or electrical level, of that point and that of the earth which we assume to be at zero potential: in precisely the same manner when we refer to the altitude of a mountain we mean the difference of level between it and the surface of the sea, which we adopt as our standard level. We assume the earth to be in a state of electrical rest, as we do the sea to be at a uniform level: and as the one is a practically unlimited reservoir of water, so the other is of electricity; small additions or subtractions of fluid having no perceptible effect in altering the level in either case. The potential of a body may therefore be defined as "the excess or defect of its potential above or below that of the earth." In the same way temperatures are measured by comparison with an arbitrary standard, the temperature of freezing water. The zero of our thermometers does not stand for an "absence of heat," but for a definite degree of heat, conveniently chosen, which serves as a point of comparison. In the same way the earth is not supposed to be devoid of electrification; but the degree of its electrification is chosen as a point of comparison. As then we have levels above and below that of the sea, and temperatures above and below that of freezing water, so we have potentials above and below the earth's potential; the former are called positive potentials, the latter negative potentials. When, therefore, we say that a point is at a positive potential, we imply that if connected with the earth, electricity will flow from it to the earth; and, conversely, that if a point is at a negative potential, electricity will, under the same circumstances, flow from the earth to the point.

It will be observed that, so far, the expression potential, "positive or negative potential of a body" coincides with the statement that it is charged with "positive" or "negative" electricity. But it frequently happens that the potentials of two points differ, though both are at a higher, or at a lower, potential than the earth's potential; we then speak of them as being mutually positive and negative, though both are, absolutely speaking, positive, or both negative. Such a case does not constitute an exception to the general law that the flow of electricity always takes place from the higher to the lower potential; from the more positive to the less positive, if both are above, from the less negative to the more negative, if both are below the earth's potential. The same thing occurs in the case of temperatures; heat flows from the warmer to the colder point, whether both be above, or below zero—just as between two cisterns at different levels, but both above or both below that of the sea, the stream flows from the higher to the lower.

CURRENT.

If we unite two insulated conducting bodies electrified to different potentials, by means of a wire for instance, there occurs almost instantaneously a change in their condition in consequence of which they are found to be both at the same potential. We express this phenomenon by saying that there has been a flow, or current of electricity through the wire from the body at the higher to that at the lower potential.—If we adopt means to prevent the equalisation of the potentials, or rather if we restore the difference of potentials to the two conductors, as fast as it is equalised, the current will be kept up in the wire, and become continuous as long as the difference of potentials is kept up.

EXPERIMENT I. Take a piece of commercial zinc and dip it in dilute sulphuric acid (1 : 4 to 1 : 10). The metal is attacked by the acid, hydrogen rapidly given off, and sulphate of zinc formed. Repeat the experiment with distilled and homogeneous zinc—no such action takes place. A few bubbles of hydrogen are evolved, which remain attached to the metal and protect it against any further chemical action.

Now take again the piece of impure zinc, and moisten it in the dilute acid, and rub into it a few drops of mercury until its whole

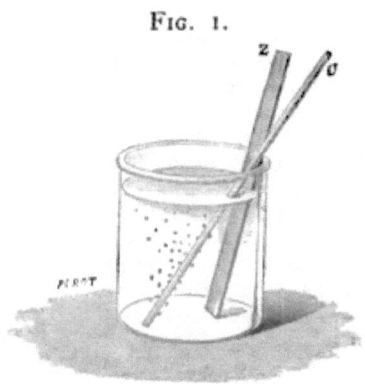

Fig. 1.

surface is perfectly amalgamated. Dip it into the acid—it behaves as the piece of chemically pure zinc, and is not dissolved.

Place in the vessel, or "cell" as it is technically called, a silver, platinum, copper or iron wire, or a rod of carbon, and let it touch the zinc outside as shown in fig. 1; or connect the two metals with a wire as in fig. 2. Hydrogen is given off again, not at the zinc, however, but at the other metal (fig. 1.), though it is obviously the zinc which is attacked by the acid.

EXPERIMENT II. Fix the amalgamated zinc and the other metal (or carbon) to the connecting wires of an electrometer, and immerse them in the liquid. The instrument shows the existence of a difference of potential between the wires. The wire attached to the zinc is negative with reference to the other, in other words its electrical level is lower. This difference differs according to the metals used. Any two metals under the same conditions give rise to it; and the following list is so arranged that, of any two chosen for the experiment, the first imparts negative potential to the electrometer, the difference of potential being the greater, the wider apart the metals occur in the list:—zinc, tin, iron, lead, copper, silver, gold, platinum, (carbon).

EXPERIMENT III. Suspend a piece of magnetised needle to a fine thread, and, when at rest bring it near the wire connecting the two metals (disposed so as to have the wire and the needle parallel). The needle is deflected.

EXPERIMENT IV. Apply the free extremities of the zinc and silver immersed in the acid, to the tongue; a slight sensation of pricking and a peculiar taste are experienced. Apply them also to a piece of blotting paper dipped in a solution of starch and iodide of potassium; a blue spot appears at the point of contact of the paper with the silver, showing the liberation of iodine.

The various phenomena observed: (chemical changes in the vessel, physiological, chemical and magnetic effects) between the extremities of the metals outside—depend upon the current of electricity generated when these extremities are connected together directly or indirectly. Now it is easy to ascertain that they are not instantaneous; they persist for a more or less considerable time. Hence we conclude that when two metals are dipped into a liquid, the difference of potential observed between them, persists even after they have been brought in connection with one another. We have seen that the wire attached to the zinc is electrified negatively, with reference to the wire attached to the other metal; hence the name of negative and positive poles given to the terminals of the two metals respectively. The flow takes place from the positive pole to the negative, as if the electricity came up from the cell at the positive, and went down into it at the negative; for this reason the greek words *anode* and *kathode* are often used to denote the conductors attached to positive and negative pole (ἀνά, upwards; κατά, downwards; and ὁδός, path).°

When the two metals are placed in mutual contact, or in contact with conducting substances, we say that "the circuit is closed," implying by this expression that the current of electricity does not simply flow from positive to negative pole outside the cell, but that within the cell itself the current flows so as to complete a circular course. The direction of this assumed internal current must, therefore, be opposed to that of the external; that is, must be from the zinc to the copper as shown by the arrow in fig. 2. The zinc and copper have therefore been called positive and negative electrodes respectively; the rule being that of any two metals, the one which is the more actively attacked by the liquid in the cell is positive to the other. We shall see presently how it is that the positive metal carries the negative pole of the cell. (*See* Theory of the Galvanic Cell, p. 9.)

ELECTROMOTIVE FORCE.

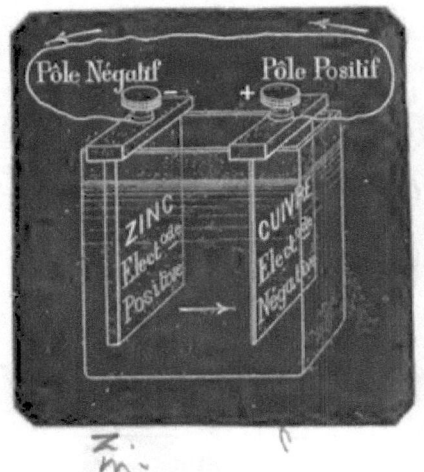

Fig. 2.

The name galvanic (or voltaic) element has been given to an arrangement such as the one just described: two metals (or carbon and a metal) immersed in a liquid, which has a more powerful chemical action upon the one than the other. A galvanic element, or cell, may therefore be described as a machine, which by the expenditure of work (chemical changes at the expense of a metal), keeps two bodies at different potentials, and thereby yields a constant flow of electricity through a conductor connecting these two bodies.

* Anode and Kathode are commonly mispronounced by making the O long, as they are misspelt by having an E for termination. They should, like their congeners method, period, etc., be spelt ánod and káthod, the accent falling upon the first syllable. I have not ventured to innovate here, as the mistake is a time honoured one. But the matter is different when we come to their derivatives. In electro-therapeutical writings anódal and kathódal have crept into use instead of anódic and kathódic. What should we think of a writer speaking of methōdal habits and periōdal returns?—There is a distinct advantage of retaining the k in kathode, for as we shall see in the chapter on physiology and diagnosis, we have to use abbreviations for kathodic closure, and closure contraction, KC, and CC, which would be ambiguous if C stood for kathodic.

A frictional machine, an induction coil, a thermo-electric couple are likewise arrangements for producing a difference of potentials between two points—this difference being manifested at the poles, the one exhibiting a certain degree of positive, the other an equal degree of negative, potential. This property of creating difference of potential depends upon an unknown cause at work within the cell, to which the name of *electromotive force* has been given. We may, roughly, compare the cell to a steam engine. As we speak of the locomotive power of the one as a fixed quantity, as long as its supply of fuel and water remains constant, so we may speak of the electromotive force of the other: the electromotive force of the cell depends upon the metals and liquids brought into contact, and remains constant, as long as these are unaltered. But here the analogy ends; and it is a point of paramount importance, to thoroughly grasp the idea that the electromotive force of a cell depends upon the nature of its constituents, and upon their nature only. The size of the cell, that is, the quantity of the metals and liquids used in its manufacture, leaves its electromotive force absolutely unaltered. Nor must electromotive force be understood to be a force like heat, light, or electricity. It is a property of matter, not a form of motion. It is a "preparedness for doing work" arising from the fact of certain substances being brought together, a condition of the evolution of electric force. *Electric* force moves *matter;* *electromotive* force is an abstraction, an imaginary force moving the abstract, imaginary fluid, *electricity*.

EXPERIMENT V. Take a piece of amalgamated zinc, and a piece of copper or carbon, and immerse them in acidulated water. Attach their upper portion, by means of two wires, to the terminals of a galvanometer. The needle is deflected. Introduce a second cell (of exactly the same composition) in the circuit, as in fig. 3, so

FIG. 3.

that the positive pole of the one cell be connected with the negative of the other. The deflection is increased. The electromotive forces of cells connected to one another *in series*, are added arithmetically.

EXPERIMENT VI. Introduce the second cell in the circuit in opposition to the first, that is, with its poles reversed; in other words, with its positive pole attached to the positive pole of the other cell, or (as in fig. 4) with its negative to the negative. The galvano-

FIG. 4.

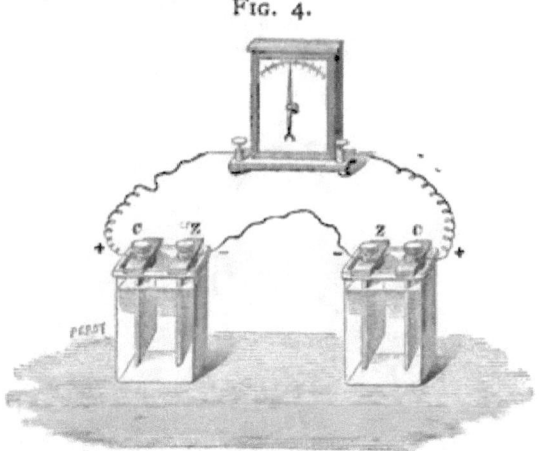

meter remains at 0. No current passes, because the electromotive forces of the two cells neutralise one another.

EXPERIMENT VII. Take a zinc and copper cell and observe the amount and direction of the deflection it produces on the galvano-

FIG. 5.

meter. Introduce in the circuit and in opposition to it, a similar cell in which the zinc has been replaced with iron. A deflection is still observed on the galvanometer (fig. 5) in the same sense as that given previously, but not so large. The electromotive forces of the two cells are opposed; but the combination of the zinc with the copper, (or carbon) creates a greater difference of potentials, than that of the iron with the same body. Hence after neutralising the electromotive force of the iron cell, the zinc cell has electromotive force left to keep up a current through the circuit.

EXPERIMENT VIII. Oppose to a cell another cell of the same nature, but twice or any number of times larger (fig. 6). The

FIG. 6.

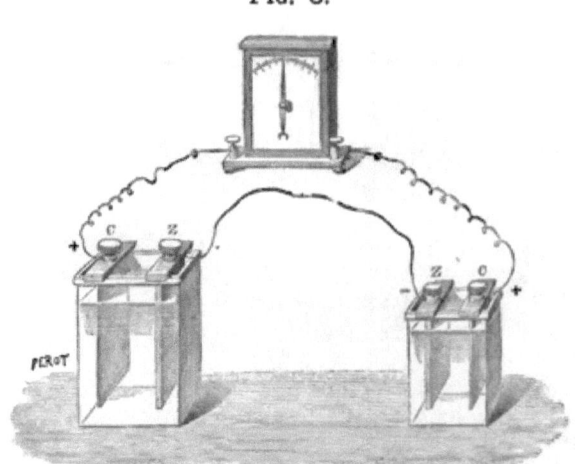

needle remains steady; no current passes. The electromotive forces of the cells are opposed to one another, and they neutralise one another because they are equal, depending upon the *nature* of the substances employed in the cell, and not upon their *quantity*.

EXPERIMENT IX. If instead of a single large cell, we oppose to our cell two (or more) cells arranged *in surface*, that is to say, side by side, and zinc attached to zinc, and copper to copper, as in fig. 7, the galvanometer shows that no current passes. Two or more cells arranged in surface, are equivalent to a single cell of the same composition, but of such a size as to have metal surfaces equal to the sum of the surfaces of the two or more cells.

For all practical purposes we can thus take electromotive force and difference of potential as synonymous; the latter may be considered as the external manifestation of the former. The two poles of a battery are comparable to two cisterns placed at different levels; connect these by means of a pipe, and a flow will take place from the higher into the lower cistern. The electromotive force of the battery corresponds to the pressure, which urges the

Fig. 7.

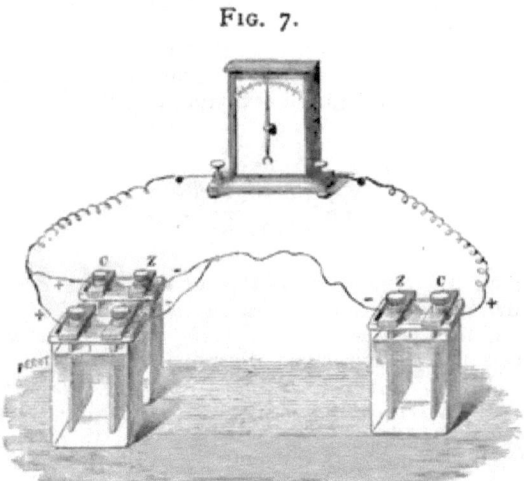

water from the higher into the lower cistern and which is due to the difference of level.

Theory of the galvanic cell. What is the part played in the generation of a current of electricity (1) by the contact direct or indirect of the metals with one another, and with the liquid which bathes them; (2) by the chemical action occurring in the cell? The answer to this question gave rise to a lively controversy; Volta had from the beginning maintained what is called the "contact theory;" Faraday, on the principle of the conservation of energy showed that in order to create force there must be an expenditure of energy, in other words that work must be done, and that contact alone could not account for the generation of the current. Hence he opposed to Volta's his own "chemical" theory, which soon generally prevailed. Lately, however, the application of finer methods has shown that both theories are right, in-so-far that whilst contact is necessary for the creation of a difference of potential, chemical action is necessary to keep up that difference of potential when the two metals are connected by means of a wire.[*] Two metals, say zinc and copper, partially immersed in a liquid display next to no difference of potential. The common error of supposing that they do, arises from the fact that the experiment just described (experiment II) was wrongly interpreted. In order to manifest difference of potential the two poles of the cell must either be in direct or indirect contact with one another (fig. 1, and experiment IV), or one of them be in contact with a wire of a differ-

---

[*] Hence the quantity of chemical change (of zinc consumed in the ordinary cell) is exactly proportional to the rapidity with which potential tends to be equalized, in other words with the quantity of electricity flowing in a unit of time from the higher to the lower potential (from anode to kathode).

ent metal from itself. In experiment II for instance, the electrometer is described as being connected with the poles by means of copper wires; it was the junction of that metal with the zinc which produced the difference of potential. The zinc, liquid, and copper were all positive, the wire alone was negative.*

RESISTANCE.

Returning to our example of the two cisterns connected by a pipe, we know as an elementary fact that the flow of water is the stronger, first, the greater the difference of level between the two cisterns; second, the larger the diameter of the pipe.† Suppose now the two poles of a cell to be connected by means of a copper wire, the same law will apply; the strength of the current, that is the quantity of electricity flowing in a given time, is determined by the electromotive force in the cell,‡ and by the diameter of the wire. The strength of a current grows with the diameter of the conductors in circuit, as well as with the difference of potential, or electromotive force of the cell.

EXPERIMENT X. Arranging one or more cells, and a galvanometer as shown in fig. 3, introduce into the circuit between the cell and the galvanometer, equal lengths of copper wire of $\frac{1}{4}$, $\frac{1}{2}$, 1 millimetre in diameter respectively. You will find that the deflection of the needle (*i.e.* the quantity of electricity passing), increases with every increase in the diameter of the wire.

EXPERIMENT XI. Instead of using the three wires of the same length but of unequal diameter, as in the last experiment, introduce into the same circuit three wires of the same diameter, say 1 millimetre but of unequal lengths, say in the proportion 1, 2, 4. You will find that the needle indicates less current as the length of the wire increases. The wire whose length is represented by 2, will be found to reduce the deflection to the same extent as did the wire of $\frac{1}{2}$ millimetre diameter in the former experiment: and the wire of four times the length of the first, will reduce it to the same point as did the wire whose diameter was $\frac{1}{4}$ mm.

EXPERIMENT XII. Repeat experiments IX, X, XI, with similar wires made of iron, and again of German silver. You will find that the relations between the series of deflections obtained in each case remain the same, but that their absolute values differ. That is, that the deflection given when copper wire is used is always larger than is the case with iron wire of the same length and diameter; and that the deflection is greater under the same conditions when either copper or iron wire is used than when the wire is made of German silver.

---

* See Fleeming Jenkin's "*Text-book of Electricity*," chap. ii.
† Neglecting the effects of friction.
‡ Assuming the internal resistance to be **negligible**.

NOTE. Experiments with wires are beyond the reach of all those who do not possess a physical laboratory at their disposition. But all the leading facts may be illustrated with an ordinary medical battery and galvanometer, and a piece of string moistened 1° with pure water, 2° with water containing traces of common salt, 3° with a concentrated solution of salt. In each case apply the clean ends of the battery wires to the extremities of the string, 1° fully extended, 2° folded double, 3° folded quadruple.

You will find in each case that the current (*i.e.* galvanometric deflection) increases with the increase of diameter of the conducting medium, the string (experiment X); and that the current is very feeble indeed when the string is moistened with pure water, stronger when it is moistened with slightly salt water, strongest when moistened with saturated salt solution, (experiment XII). Experiment XI may be repeated by placing the electrodes at, say, 1, 2, 4 inches from one another on the fully extended string, when you will find that the deflection diminishes with the increasing distance between them.

It is found, however, more convenient to speak, not of conductivities, but of resistances. Resistance is the reciprocal of conductivity, and the one easily reducible to the other if we remember, that the reciprocal $\frac{1}{R}$ denotes the conductivity of a body of which the resistance is represented by R.

It is obvious that we may say indifferently, speaking of the two wires (or metals) A and B, that the conductivity of A is 100 times better than that of B, or that its resistance is 100 times less than ($\frac{1}{100}$) that of B, and conversely that the resistance of B is 100 times that of A, or that its conductivity is $\frac{1}{100}$ that of A.—All bodies, even the best conductors, offer some resistance to the passage of the current. This proposition is self evident when we consider that electricity like all other forces of nature, is a mode of motion, and that all matter offers a certain amount of resistant inertia to its being put in motion. The best conductor of electricity, as of heat, is the body whose molecules vibrate with least difficulty when exposed to the action of the force.

Measurement of currents under various conditions proves that:—

1. The resistance of a wire or other conductor is proportional to its length, (experiment XI).

2. This resistance is inversely proportional to, in other words, diminishes with every increase of, the area of its own section, (Exp. X).

3. The resistance of a conductor of given length and diameter depends upon the material of which it is made, *i.e.* upon the *specific resistance* of the material, (experiment XII).

Let us carry these new factors of the case into our comparison of electrical conductors with water pipes. Since all electrical conductors offer some resistance to the current, we must suppose the pipes not to offer an absolutely free passage to the water, but to have their lumen occupied by a porous substance, or to be packed with bits of sponge or pumice. In such pipes, the diameter in each case will influence the amount of water transmitted under uniform pressure. But two other elements will enter into the determination of that amount:—First, the tightness of the packing, or density of the porous material; second, the length of the pipe. The former constitutes the "specific resistance" of the pipe to the passage of water. The second obviously follows the consideration that in such a pipe two feet long, twice as much friction, or resistance will have to be overcome as in one half that length.

These considerations will assist us in realising the property of bodies considered as electrical conductors: some transmit the current freely, some with greater difficulty, some very sparingly indeed. Metals are the best conductors. Even those which conduct badly among them, such as mercury or German silver, do so far better than non-metallic bodies. Carbon conducts fairly well. Acids and solutions of salts have much less resistance than pure water. Bodies such as resins, India-rubber, silk, glass, dry air conduct so badly as to deserve the name of "insulators." The human body is a conductor, its chief constituent being a saline fluid; the dry epidermis, however, almost acts as an insulator, but becomes a conductor in proportion as it is soaked with a conducting liquid.

The following are the specific resistances of various bodies:— Silver, 1·6; copper, 1·64; iron, 9·8; German silver, 21; mercury, 96; pure water, 7×10^{10}; acidulated water (about 1 $H_2 SO_4$ in 10), 3×10^{10}; glass, 2×10^{16}; gutta percha, $3·5 \times 10^{23}$. These numbers teach us that in a circuit containing wires and conducting liquids (such as those of the human body) the resistance of the former are quite negligible, at least under ordinary circumstances.

Let us now put the two ideas of electromotive force and resistance together in the shape of a formula. The *Current* of electricity will be directly proportional to the *Electromotive* force (and to the conductivity that is to say) inversely proportional to the *Resistance* in the circuit. Or mathematically expressed, $C = \frac{E}{R}$. This is *Ohm's law*, the foundation-stone of electrical science.

Let us consider R a little more closely. It represents the resistance in the whole circuit. In travelling from the positive to the negative pole of a cell, the current encounters the resistance of the wire, human body, etc., which are included in that part of the circuit. This we call the *external* resistance. But inside the cell, what takes place? The current, in order to complete the circuit, must flow from the positive to the negative metal, and there also it encounters a resistance, viz., that of the liquids and diaphragms interposed. This we call the *internal* resistance, and our equation $C = \frac{E}{R}$ becomes $C = \frac{E}{R' + R''}$ where R' is the resistance in the battery, and R'' that outside. We shall see that R' is often a most important element in electrical problems, and one which is too often wholly lost sight of by those who have not thoroughly mastered the essentials of electrophysics.

ELECTRICAL UNITS.

Before proceeding any further, we must pause to consider what actual quantities the letters C, E, R represent in the equation just given. C is equal to 1 when $E = R$; it is evident, therefore, that we may call a current of unit strength, that which circulates through a resistance of one unit, generated by an electromotive force of one unit.

Until lately the unit of potentials was the electromotive force of a Daniel's cell, whilst the unit of resistance was arbitrarily chosen, as that of a piece of copper wire, or column of mercury, of given length and diameter. Now when we remember that the electromotive force of a cell is a quantity subjected to variations, depending upon changes in its chemical constituents, it is obvious that no very definite results could be obtained from the use of such a standard. As the industrial uses of electricity spread, however, the necessity of fixed standards of measurement become more and more felt; and after protracted labours, and innumerable experiments with most delicate and complex apparatus, the committee appointed for that purpose by the British Association promulgated a system of electrical units, based upon the absolute units elaborated by Gauss, Weber and other physicists. It would be quite beyond the scope of these pages, to refer with more detail to the methods used in the determination of electrical quantities expressd in terms of absolute units. Suffice it to say that the units chosen for E, R, and C, are multiples of the absolute, or C. G. S.[*] unit; and that these multiples were fixed so as to correspond to quantities convenient in practice.

The Volt or unit of electromotive force is a little less than the electromotive force of a freshly charged Daniell's cell (1 Volt = 10^8 C. G. S.). The Ohm, or unit of resistance is to the Siemens' unit as 1 to ·9705 (10^9 C.G.S.), hence 20 Ohms = 21 Siemens' (nearly). The Ampere, or unit of current strength is thus one tenth of an absolute electro-magnetic unit of quantity, and represents a current conveying the unit quantity of electricity, that is one weber, during one second.[†]

It is a great defect in the old-fashioned text-books of electricity, as well as in most electro-therapeutical treatises, that they convey no definite idea by the enunciation of Ohm's law, on account of their leaving the terms electromotive force, etc., in the state of mere abstractions, without embodiment into actual magnitudes.

We speak of a battery of so many *Volts* as we would of an engine of so many horse-power. In both cases these expressions designate the capacity of the machine to overcome resistance. We may consider the *Ohm* the unit of resistance in very much the same light as we do the metre the unit of length. To the practical electrician, the Ohm is simply a standard coil of copper wire 1 m.m. in diameter, and 48·5 m. (nearly) in length; or else a column of mercury 1 sq. m.m. in area, 1·05 metre in height. A little practice only is required to realise the facts conveyed by these terms; and we shall be amply repaid for our trouble by the clearer conception of electrical phenomena, and the possibility of recording our observations

---

[*] Centimetre, gramme, second.

[†] The unit current strength used to be called the weber, until the late international congress of electricity altered its name, on adopting the B.A. system of units. The use of the term weber is restricted to the unit of quantity; the ampere denotes the same quantity, but includes the time factor implied in the word current.

in accurate and universally understood language. It is mere waste of time to speak in vague and often erroneous terms of the "high tension" of batteries of the "infinite resistance" of the human body. Much useless controversy has been expended over matters which could have been explained in an instant by reference to the facts of the case. Thus, for instance, on finding that the current from 35 sulphate of mercury elements vesicates the skin in a very short time, whilst the current from 45 sulphate of copper elements of special construction can be borne without discomfort for a long time, it has been concluded that the current from the latter has qualities which especially suit them for therapeutical purposes—whereas the facts are these: the 35 mercury elements have an electro-motive force of about 52 volts, and an internal resistance of 100 ohms. The 45 copper elements have 48 volts E.M.F., and 1500 ohms R'. What happens then is this:—in the first case the current flowing through the human body (whose resistance is, say, 2500 ohms).

$$\frac{52}{100 + 2500} = \cdot 02. \quad \text{In the second} \quad \frac{48}{1500 + 2500} = \cdot 012.$$

In other words the current in the first instance is nearly twice as strong as in the second. We shall have the opportunity further on of showing that the want of familiarity with the conception of electrical resistance is at the root of much of the obscurities which still shroud the medical application of the currents, and has caused many of the various mistakes with which electro-therapeutical writings swarm.

As practical men, having to do with electricity considered as a fluid to be administered in definite quantities, it is of paramount importance that we should have a thoroughly clear understanding of the idea expressed by the words *strength of the current*,[*] and its measurement.

A current of a given strength, be it a current of water, or electricity, or of any other "fluid," performs a certain amount of work in a given time. The current of water causes a wheel to revolve so many times in a minute; the current of electricity decomposes in a second, so many grammes of water, salts or animal tissue, heats a wire to such a temperature, magnetises a piece of iron to such an intensity, exerts (within physiological limits) such an action upon the organism. With every change in the strength of the current, its mechanical, physical, chemical or physiological effects display a proportionate change in their intensity. It matters not what be the source of the current; it may be chemical work as in the case of the voltaic pile, or mechanical work, as in Gramme's machine; its effects are the same, its strength being the same. What constitutes the strength of a current, that is, its capacity for work, is the quantity of fluid conveyed in a unit of time through any sectional area of the channel. A current of unit strength, then, is one which conveys a unit of quantity in a unit of time.

---

[*] Current strength is the term now used instead of "intensity of current," a useless and misleading expression imitated from the French.

It is a current which conveys, in the case of water, one litre in in the second; in that of electricity, one weber in the second.

The *ampere* is the unit of current strength; and by Ohm's law is that of a current, furnished by an electromotive force of one volt through one ohm, or, generally expressed, by n volts through n ohms $\left(\frac{n}{n} = 1\right)$. This magnitude becomes realised to our minds when we embody it in the form of the visible work it yields. A current of one ampere sets free, in one second, 114·6 cubic millimetres of hydrogen and 57·3 c.mm. of oxygen from the electrolysis of water, (at 0°. and under 760 mm. pressure). It deflects a given galvanometer-needle to a certain number of degrees. It produces on a given portion of tissue, a definite physiological effect.

It is clear then that when we speak of currents of ·01 1, 10 amperes, &c., we designate currents conveying ·01, 1, 10 webers in the second; or conversely, conveying 1 weber in 100, 1, ·1 second respectively—and producing an effect proportional to the quantity of electricity they convey.

The strength of a current is measured in practice by observing the amount of work, chemical or mechanical, it does. The voltameter enables us to measure it by the quantity of gas set free by the electrolysis of water. Since we know that one weber of electricity sets free 171·9 cubic millimetres of gas, we can readily calculate the strength of a given electrical current by noting the quantity of gas evolved in a given time. The galvanometer enables us to do the same by the amount of deflection of the magnetic needle; this amount bearing a certain proportion to the force exerted by the current in overcoming the directive influence of the earth upon the needle.

We shall return to these points when discussing the measurement of current strengths applied to the human body, and show the advantages accruing from the adoption of the thousandth of an ampere as a working unit, and calling it a milliampere. This unit exactly suits the requirements of medical practice, for a current of one milliampere is that given by three Daniell's through the average resistance of the human body—that is, through parts of medium resistance, and through medium sized electrodes. Again, no system of measurement is likely to be adopted unless its unit has a convenient name; and it is obviously easier to record strengths of 1, 5, or 15 m.a. than strengths of ·001, ·005, or ·015 of an ampere. We have a parallel instance in the use of the metre and millimetre. Experience shows that, applied with medium sized electrodes, medical currents range between 1 and 20 milliamperes. Hence we may conveniently designate such currents as "very weak," "weak," "moderate," and "strong," according as they range from 1-5, 5-10, 10-15, 15-20 milliamperes respectively. "Very strong" currents would in this classification range between 20 and 40 m.a., but for reasons which will become clear hereafter, presuppose the use of larger electrodes.

ON THE GRAPHICAL METHOD OF REPRESENTING ELECTRICAL PHENOMENA.

The reader who wishes to obtain a practical grasp of the phenomena of current electricity, cannot do better than accustom himself to represent the relations between electromotive force, resistance, and current strength by the following graphical method. Let the length OC represent the total resistance of the circuit of which OB

Fig. 8.

is equal to the (internal) resistance (R') of the battery, and BC to the (external) resistance (R'') of the conductor interposed. Draw OA at right angles to OC representing the electromotive force (E) of the battery. Join AC. The current strength will be represented by the *slope* of AC, that is, by the tangent of the angle ACO, which we know is the geometrical expression for the proportion $\frac{AO}{OC}$ (that is $\frac{E}{R'+R''}$).

Let OA be 15 millimetres long, representing the electromotive force of 8 Bunsen's cells (15 volts), and OB be 10 m.m. long, representing the internal resistance of the battery (10 ohms). Join AB. The strongest current we can obtain with our battery, is when its poles are directly connected, and then our current = tan ABO. If the external resistance be equal to 5 ohms (*e.g.*, a piece of platinum wire) prolong OB to OC', making BC' = 5 millimetres. Then OC' being equal to AO, the angle AC'O has 45 degrees, and our current is equal to 1 (tan 45); for $\frac{15 \text{ volts}}{15 \text{ ohms}}$ = 1 Ampere. With every increase of our external resistance, OC, OC$_2$, OC$_3$, the slope of AC$_2$ AC$_3$ goes on diminishing, until it vanishes altogether; when we have AY parallel to AC prolonged to infinity. An infinite resistance being interposed, no electricity can flow; that is, the circuit is broken.

Now draw BD parallel to AO. The length of BD represents the difference of potential between the poles of the battery when the current flows through the external resistance BC, that is to say, the "tension" of the battery, or again, the electromotive force which is effective in maintaining the current through BC, after overcoming the internal resistance OB. With every diminution of BC, (BC') BD will diminish (BD'); until BD vanishes altogether (B) when there is no external resistance, the poles being in contact.

With every increase of BC, (BC$_2$, BC$_3$) BD increases until it becomes By = AO, when the circuit is broken. The potential of any point in a circuit is therefore equal to the ordinate of that point.

Let us apply these principles to a concrete example: To find the electromotive force and internal resistance of a given battery, with a given tangent galvanometer, and a given external resistance. First, connect the poles of the battery to the terminals of the galvanometer; let the deviation of the needle = angle x. Next introduce into the circuit the given resistance R, and let the deflection be then = angle x'. Take a straight line OA, and make the angle OAP = x. Produce the line OA to A, and let AA' represent R; make the angle OA'P' = x'. Let AP and A'P' meet, as they must, at a point Q. From Q let fall QO perpendicular to AO produced if necessary. Then OA = internal resistance of the battery + known resistance of the galvanometer, and OQ = electromotive force of the battery, in terms of that electromotive force taken as unit, which if it acted in a circuit of unit resistance, would generate a current capable of causing a deviation 45° upon the galvanometer employed. That is, if every inch, millimetre, or any other unit of length adopted for AA' denotes one ohm, the length of OQ expressed in terms of the same unit denotes the electromotive force of the battery in volts, and the length OA denotes the internal resistance of the battery, plus the resistance of the galvanometer; for $\frac{1 \text{ volt}}{1 \text{ ohm}}$ = 1; and tan 1 = angle 45°.

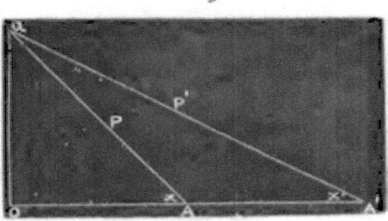

Fig. 9.

Let A and B be two points in a circuit, the two poles of a battery for instance, and the length of AB represent the resistance of an uniform conductor interposed. Let AC be made equal to the electromotive force acting at A (*i.e.*, the potential of A) and BD that at B, the potentials of these two points being positive and negative respectively. Then the slope of CD, that is the tangent of the angle AOC represents the strength of the current; and the potentials of any points *abcd* along AB are equal to the ordinates of these points (drawn in the diagram as full lines between AB and CD). The point O, which is at zero (the earth's) potential, occurs at the middle point of AB.

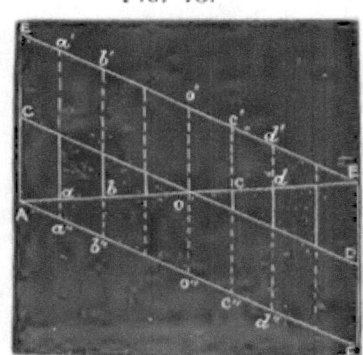

Fig. 10.

By means of an "earth wire," that is of a wire connected with the earth (this is usually done by attaching the wire to a gas or water pipe), any other point in the circuit may be brought to zero potential. Suppose we so connect the negative pole B of the battery: it is then at zero. But the electromotive force of the battery remains unaltered, and the difference of potential between A and B is still equal to AC + BD, that is equal to AE. The current remains unaltered for ABE = AOC; but there occurs a redistribution of potentials along the whole of AB; the potential of every point A, a, b, O, c, d, B, being raised by the value of BD. Hence the ordinate AE, aa' bb' Oo', etc., represent these potentials which are all positive. The opposite occurs if the point A is connected to earth. The potentials aa' bb' Oo' are now all negative. Any point intermediate between A and B being so connected, a corresponding redistribution of potentials occurs: in order to represent the state of things when the point a, for instance, is connected to earth, we have simply to draw a line through a parallel to CD; the potentials of A, b, c, d, B would then be represented by the ordinates of these points.

MEASUREMENT OF ELECTROMOTIVE FORCE.

ELECTROMOTIVE force may be measured statically by observing the potential (that is, the "tension" of the older authors) of the point at which it acts. For this purpose one of Sir W. Thomson's electrometers is used. These instruments are of two very distinct forms: the 'portable' and the 'quadrant' electrometer. The action of each depends on measurement of the attraction between two planes, one of which is electrified to a constant potential, and the other brought to that which is to be measured. In both forms the potential of the electrified plane is kept tolerably constant by being connected with a Leyden jar of considerable capacity, formed by the case of the instrument. The ordinary portable instrument will measure a difference of potential, not less than that of about 1 Daniell's cell or Volt. The quadrant electrometer is much more sensitive, and will indicate a difference of potential of $\frac{1}{100}$ of a Volt.

The electromotive force of a cell or battery may be estimated by comparison with that of some known constant element taken as unit. The Daniell is usually so chosen. A circuit is formed containing a rheostat, a galvanometer and the known electromotive force E. A convenient deflection having been obtained by intercalation of the necessary resistance, the second electromotive force E' is substituted for the first, and the current brought to the same amount by means of the rheostat. Let R be the total resistance in the first case, and r that in the second; then we have $\frac{E}{E'} = \frac{R}{r}$.

By using a sensitive galvanometer and a high rheostat resistance, the resistance of the battery and galvanometer may be neglected. Other methods exist, but need not be explained here.

In order to measure in absolute units the electromotive force of a given cell or battery, we may make two measurements on a tangent galvanometer with different rheostat resistances. Suppose a cell gives with R and r ohms, currents of C and C' amperes respectively; then its electromotive force $E = CC' \frac{r-R}{C-C'}$.—We have already seen (p. 17) how to perform the same operation without calculation, by means of the graphical method.

The electromotive force of a battery may be estimated by a single measurement with a known resistance in circuit, upon a galvanometer divided into milliamperes.*

This is the easiest way to proceed when approximate results only are required, as would be the case in medical practice. Thus 20 freshly charged Leclanché's through 1000 ohms would give a deflection of about 30 milliamperes. Hence we conclude that their electromotive force is equal to about 30 volts (1·5 volt per cell), for 30 m.a. $= \frac{30 \text{ volts}}{1000 \text{ ohms}}$. If after a year's use the battery had lost much power, we should find that the deflection, with the same resistance and galvanometer, was much less—say 10 or 15 milliamperes. Hence we should conclude that the electromotive force had sunk to $\frac{1}{2}$ or $\frac{1}{3}$ of its former state—provided always the internal resistance of the battery to be still negligible, which is not always the case when the small medical elements are tested. These are liable to alterations, the consequence of which is an enormous increase of their internal resistance. In order to determine whether the diminution of the current strength is due to an impairment of the electromotive force or an increase of the internal resistance, we must oppose to the battery (see experiment VI, fig. 4) an equal number of freshly charged cells of the same description, (or a proportional number of cells of known electromotive force), and note whether the needle is deflected in the opposite direction. If it is not deflected we know that our battery has preserved its electromotive force, and that its diminished action arises from its augmented internal resistance. If the needle is deflected we know the electromotive force to be deficient; and we may at the same time determine this loss, by finding how many cells of the opposed fresh battery are required to neutralise the battery we are testing.

---

* A galvanometer may also be used for this purpose, whose dial has been experimentally divided so as to indicate directly the E.M.F. in volts, provided that the internal resistance of the elements measured be negligible.

MEASUREMENT OF RESISTANCES; RHEOSTATS.

WE have seen that the total resistance is made up of the internal R', *i.e.* the resistance of the battery, and the external R", *i.e.* in electrotherapeutics the resistance of the human body and electrodes.

In order to estimate the internal resistance of a cell, the simplest method is to connect its poles with the terminals of a tangent galvanometer or of one divided into milliamperes (whose resistance we assume to be negligible), and to note the deflection. This gives us the current strength C which as we know is the quotient of E by R. We now introduce into the circuit a rheostat, and put in resistances until the needle shows that C is diminished by half. It is obvious then that the additional external resistance must be equal to the internal, for since $\frac{E}{R'} = 2 \frac{E}{R'+R''}$ R' = R".

Or we may by means of the graphical method (p. 17) determine the resistance from two observations on a tangent galvanometer, as described above, and a simple geometrical construction.

If it is the resistance of a part of the human body that we wish to determine, we first note the deflection given when that part is in the circuit. We then replace it by the rheostat, and so arrange our artificial resistance, that the deflection obtained is the same as before: this resistance must be the same as that of the body, for $C = \frac{E}{R'+R''}$ in both cases, hence R" must be the same.

A Rheostat (using this term in a wide sense) is an instrument so constructed as to allow the interposition of definite resistances in a circuit. Of the numerous forms that have been devised, only two or three are of importance to the medical electrician.

Wire Rheostats, otherwise called resistance coils, consist of coils

FIG. 11.

Set of resistance coils. Measured resistances are introduced into the circuit by placing into it the coils marked 1, 2, 5, 10. S, S, binding screws for rheophores. A, B, C etc., handles for putting the resistances in and out of circuit.

of fine German silver wire. Each coil has a definite resistance of so many ohms, depending upon the length and fineness of the wire; and by a simple mechanism can be brought into the circuit. It is easy with such instruments to introduce any required resistance in a circuit, but they are rather expensive.

Liquid Rheostats consist of a column of water, either pure or containing a salt, such as sulphate of copper, in solution. The instrument is usually made of a piece of glass tubing capped at both ends with a metallic disk. To these disks are attached the rheo-

FIG. 12.

Liquid rheostat.

phores; and through one of them a metallic rod works like a piston. The rod is long enough to touch the opposite disk when fully pushed in. The current, then, encounters no sensible resistance; but as the rod is withdrawn and a layer of liquid is interposed, a resistance is created which will be directly proportional to the thickness of that layer and to the specific resistance of the liquid employed, and inversely to the diameter of the tube.

A rough but useful rheostat can be extemporised out of a piece of glass tubing, two corks, and some stout copper wire. The tube, filled with the liquid) is corked up at both ends. Through one of the corks a short piece of wire is inserted which just dips into the liquid; whilst through the other a longer piece acts as a piston. The rheophores are attached to the wires. When very high resistances are required, and the tube must be short, pure water may be used, (fig. 12.) Usually, however, a solution of sul-

FIG. 13.

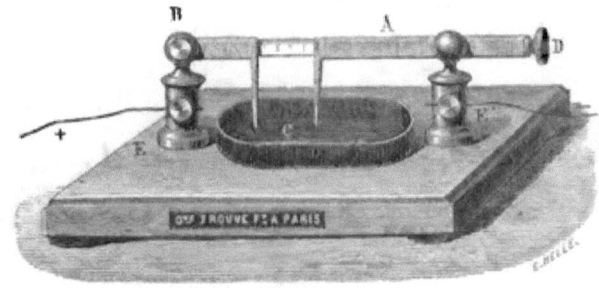

Trouvé's liquid rheostat. The rod A is regulated by means of the brass head D. The length of the column of liquid C, interposed in the circuit, is read off the measured scale. The current enters at E, passes through the pillars B, through the metallic pieces and points, and the liquid; making its exit E'. The scale is made of ivory or some such insulating material.

phate of copper is preferable as it diminishes polarisation. A column of saturated solution 1 metre high, ·5 square centimetres area, gives a resistance of about 6000 ohms. The more dilute the solution, the greater its specific resistance; hence the shorter the tube required.

Liquid rheostats should be accompanied with a scale corrected from simultaneous readings on a set of resistance coils. Even then, however, they are not reliable for accurate measurements, though amply so when used for introducing approximate resistances in a current for medical purposes. The inconveniences attending the handling of long tubes filled with liquid prevent unfortunately their ever becoming extensively adopted in electro-therapeutics.

A third kind of rheostat, which possibly may yet be found the most suitable for medical practice, though much is yet required to bring it to perfection, is that in which *powdered graphite* is the resisting material. The substance may be simply spread out in thin lines on a vulcanite board, when a resistance is to be extemporised, or reduced to a solid compound by admixture with various non-conducting bodies, such as gums, in such proportions as to yield hard cylinders of the requisite specific resistance. Attempts are being made to overcome the difficulty arising from the fact that the resistance of such cylinders is altered considerably by use; if successful, we may be hope to be placed in possession of a cheap and useful instrument.

MEASUREMENT OF CURRENT STRENGTH.

WE have seen that the expression "strength of the current" means the quantity of electric fluid conveyed in a unit of time, in other words the amount of work, chemical, mechanical, etc. done by the electricity in the circuit. Hence the work performed indicates the strength of the current flowing. Instruments for measuring currents are based upon this principle; the two kinds chiefly used being the voltameter and the galvanometer. The *voltameter* is an arrangement in which the chemical action of the current furnishes the index of its strength. Electricity in passing through water decomposes it into hydrogen and oxygen.

EXPERIMENT XIII. Take a few cells connected in series (as in all medical batteries, fig. 23), and attach to their poles two wires, the free extremities of which you plunge in a glass containing water, or acidulated water. You then observe that the immersed portions of the wires become coated with fine bubbles of gas, which presently escape in greater or less abundance through the liquid. No gas is given off from the water between the wires but only at the points at which they are in contact with the water. The quantity of gas liberated at one of the wires is much greater than that evolved at the other. The former you will find to be the wire connected with the negative pole, or kathode of the battery; and the gas escaping around it is hydrogen. The latter is the anodic or positive wire at which oxygen is given off. These gases (2 vols. of H, 1 vol. O) are the constituents of water (H_2O), which is decomposed or electrolysed by the electric current.

The weight of water electrolysed or volume of gas given off is proportional to the quantity of electricity flowing. If then, the amount of gas given off by a current in a certain time be collected and measured, we shall be able to determine the strength of the current, remembering that the unit current of one ampere gives off nearly 115 cubic millimetres of hydrogen, and 57 of oxygen, in a second, that is about 10·3 c.c. of the mixed gases in one minute.

Voltameters are made for collecting the two gases, either separately or together. They consist mainly in a vessel filled with acidulated water into which project two wires or plates of platinum to which the wires from, and to, the battery are attached. The products of electrolysis are collected in one or two glass tubes standing over these electrodes; the hydrogen is given off at the negative, the oxygen at the positive electrode, (fig. 14).

FIG. 14.

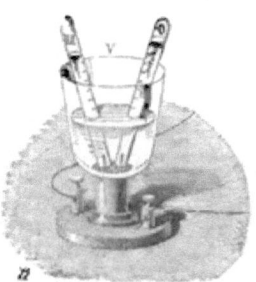

Voltameter for collecting the gases separately, H and O. The wires connecting the instrument to the battery reach the binding screws at the base. The projecting platinum electrodes, connected with the latter, are seen to project into the liquid at the bottom of the glass vessel. Two inverted tubes are so disposed as to receive the bubbles of hydrogen and oxygen given off from the electrodes, and graduated so as to allow the amount of gas collected by displacement of the liquid which filled them, to be read off at a glance.

A voltameter has recently been constructed by Gaiffe which is calculated to fulfil many of the requirements of the physician desirous of acquiring the means of estimating the strength of his battery, or of the currents he uses for therapeutical purposes. The instrument is represented in fig. 15, and is used as follows: The tubes are filled with water and the corks fixed to their respective mouths. The current led by the platinum wires or "electrodes" decomposes the water into oxygen and hydrogen, which ascend in the inner tube and collect at the top. When the inner tube is filled with gas, the inner cork is lifted out by means of the wire, the water enters into inner tube from the outer, and the instrument is ready for further measuring. Now and then a little water has to be poured into the instrument, so as to secure the e-r

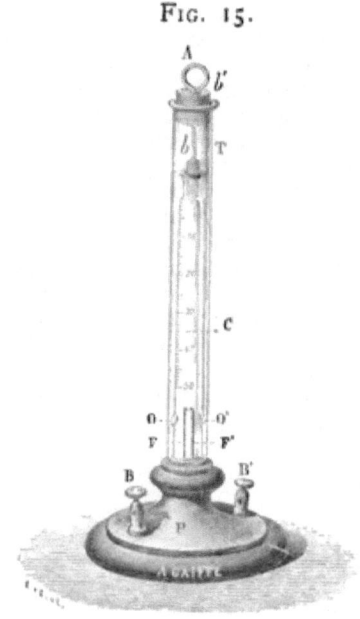

FIG. 15.

T outer tube, C inner tube, both fixed to a stand, P.—b and b' corks closing the upper extremities of the tubes. b can be lifted out of the inner tube by pulling the wire A which passes through the cork b'.

O, O' apertures in the inner tube; FF' platinum wires projecting into the inner tube and connected with the binding screws BB' to which the wires of the external circuit are attached.

The inner tube is graduated, so as to indicate the volume of gas set free in cubic millimetres.

filling of the inner tube when the cork is raised. This voltameter is a cheap, simple, and reliable instrument for estimating the strength of currents, the electromotive force of the battery, and the resistance in circuit. But its applicability is limited to cases where the current is allowed to flow for one or more minutes. For the measurement of currents of short duration, such as are used in electro-diagnosis for instance, a galvanometer is necessary. A few examples will assist the reader in realising the method of measuring electrical magnitudes by means of the voltameter.

Place the voltameter in circuit with the battery and the human body. Note the amount of gas collected in the graduated tube, during any one minute, (1st, 2nd, 3rd, ... 10th, 11th, etc. minute). Since 1 milliampere of electricity sets free 10·3 cubic millimetres of gas per minute we have, in order to know approximately the strength of the current at any time, only to divide the number of c.mm. of gas set free during one minute by ten. Or if we want to know the total amount of electricty used during a certain number of minutes, to divide the total number of c.mm. of gas set free by ten.

In order to compare the electromotive force of two batteries, place them successively in circuit with the voltameter and the same external resistance (for instance, the same portion of the human body, or better the same amount of rheostat resistance), the amounts of gas liberated in equal times are then proportional to their electromotive forces.

In order to compare the resistances of two conductors (for instance, that of an arm with that of a leg, or another arm, etc.), note the amounts of gas liberated in equal times with the one, then with the other conductor in circuit by the same number of similar cells: the resistances are inversely proportional to those amounts.

We have already seen that a wire through which a current circulates, deflects a magnetised needle when brought it its neighbourhood. This deflection occurs in a definite direction, which may be remembered by imagining yourself swimming with the electrical current, and with your face always looking towards the needle. You would then see the north pole of the needle always deflected to your left side.

A current flowing from South to North above, and a current flowing from North to South below the needle, would tend to deflect it in the same direction; and their influences upon the needle would be added. In order, therefore, to make a needle sensitive to currents otherwise too weak to deflect it visibly, the wire is arranged into a coil consisting of a great many turns, each of which acts upon the needle which it surrounds. The total effect will thus be equal to the deflecting power of one turn multiplied by the total number of turns; hence the name of 'multiplier,' given to a galvanometer provided with such a coil.

The *galvanometer* may be briefly described as an instrument consisting of a magnet freely suspended or pivoted in the centre of a graduated disk, so as to be easily deflected by the passage of a current in a coil of insulated wire, properly disposed in its neighbourhood.

Equal deflections on the same galvanometer always indicate currents of the same strength. But it must be carefully kept in mind, 1st, that the angle of deflection is not proportional to the current, and 2nd, that it differs for the same current strength from galvanometer to galvanometer, depending upon the length or number of turns of the wire, and upon the position of the coil with reference to the needle.

Sometimes the needle is placed within the coil as in fig. 16, and

Fig. 16.

Du Bois Reymond galvanometer (or multiplier) with 25,000 (or more) turns of fine wire for measuring physiological currents.

is rigidly connected with a pointer, both being suspended by means of a fine silk thread. The pointer revolves in the centre of a subdivided dial, and may consist of another magnet forming an "astatic" system of much greater sensibility. A fuller description of these and the following instruments would be out of place here; the reader is referred to the usual electrical text-books for further details.

The *sine* galvanometer is an instrument so constructed that the sines of the angles of deflection produced upon it by currents of different strengths, are proportional to these strengths. (Fig. 17).

FIG. 17.

Sine galvanometer.—A magnetic needle a, b, carrying a long pointer c, d, revolves on a pin in the centre of the dial A, A, divided into 360 degrees. B, B, is a circular frame supporting the coil of the wire t, t'. A spirit level h, serves to secure the horizontality of the instrument.

As far as this the instrument is mainly the same as the tangent galvanometer, with the exception that in the latter, the needle must be much smaller. The remaining parts consisting of a graduated circle C, and an index f, serve to determine the total angle of deflection, the sine of which is proportional to the strength of the current.

On the *tangent* galvanometer the tangents of the angles are proportional to the current strengths. None of these instruments give the absolute current strength, however; the results given by each are in terms of itself. In order to obtain measurements in absolute units, the sine or tangent must be multiplied by a constant, called the reduction factor of the instrument. For most medical purposes their size, price, and complication render them unpracticable.

The ordinary galvanometer, or rather *galvanoscope*, which is a much simpler instrument, indicates, whether any current is passing or not, whether the battery is constant and whether the resistance in circuit varies. As it obeys no known law in the deflections it gives, no calculation can transform readings obtained by it into absolute, not even into proportional, measurements. It consists simply of a coil of wire influencing a magnetic needle, revolving on a pin in the centre of a dial divided into degrees of a circle. (Fig. 18).

FIG. 18.

The needle instead of being placed in a horizontal plane may be pivoted so as to hang vertically, and the coil of wire is made to occupy a corresponding position. In some respects this is a con-

FIG. 19.

Vertical galvanoscope. The needle and coil of wire are contained in the rectangular box. The axis of the needle is prolonged externally, and carries a vertical pointer which indicates on a scale the angles of the deflections produced, by currents. In order to reduce the oscillations of the needle, its axis may be fitted with a projecting piece dipping into a vessel filled with oil and contained in the box.

venient arrangement; the needle takes less time to settle when the current is made, and can be more readily observed from a distance, but its indications are subject to variations which render them less accurate. In order to give uniform results the vertical galvanoscope requires that its needle should always be magnetised with the same intensity, an almost impossible condition to fulfil, since all magnets lose their force at a more or less rapid rate, or are altered by the passage of currents in their neighbourhood. Still for medical purposes the vertical galvanoscope is by no means to be rejected; periodical remagnetisation of the needle suffice to ensure all the accuracy required in practice. From the position of the needle, the variations in the horizontal magnetic force of the earth cease to have any influence, and the graduation of the vertical galvanoscope holds good, for all times and all points of the earth's surface, provided always that the magnetisation of the needle remain uniform. The work done by the current in deviating the needle of the vertical galvanometer, consists in overcoming the force of gravity acting on the needle, instead of the directive magnetic force of the earth, as is the case in the horizontal galvanometer.

It has been the merit of M. Gaiffe of Paris to convert the galvanoscope into a really useful and accurate instrument for the purposes of the physician. Instead of dividing the dial into degrees, he graduated it into subdivisions of the ampere. This most valuable improvement introduced in 1873 seems not to have attracted the attention it deserved until 1877 (see the *Lancet*, vol. 1), and 1878 (see first edition of this book pp. 5, and 21), when I insisted upon the introduction into electro-therapeutics of a system of measurement, and the graduation of galvanometers in absolute units. Gaiffe's ori-

FIG. 20.

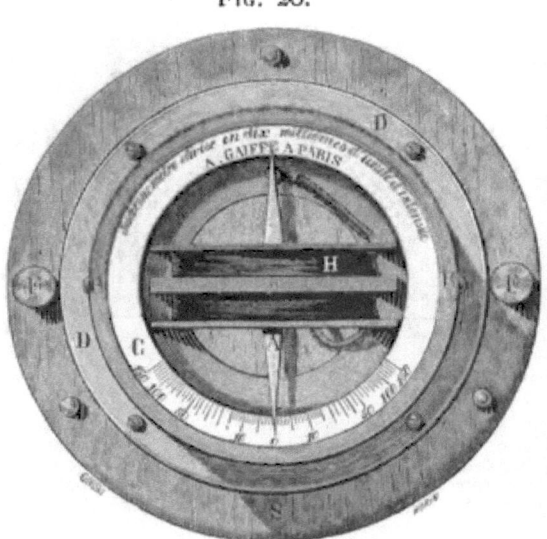

ginal instruments were subdivided into "ten-thousandths of a unit of intensity." This cumbrous terminology, as well as the fact that this subdivision was unnecessarily small made me propose the adoption of the "milliampere"—then called milliweber—as medical unit of electrical quantity or current strength.

The grounds upon which I urged the adoption of the milliampere graduation were that, apart from the necessity of having a convenient name to designate the standard quantity, the milliampere is the most practical unit of measurement. For its multiples correspond to the strength of the currents used in medical applications. The resistance of the parts of the human body included between electrodes of medium size and applied to the spots commonly selected or indicated in the more ordinary cases of treatment averaging in ordinary individuals from two to four thousand ohms, a current of one milliampere, would be yielded by 2 to 4 Daniell's cells, and is about the weakest ever used therapeutically or diagnostically. Likewise its multiples 5, 10, 20 express currents yielded by 10 to 30 or 80 cells under similar conditions. It is apparent, therefore, that in the milliampere we have a convenient unit by which to express the electrical doses, as it were, administered to patients; or, the current strengths necessary to obtain muscular contractions in electro-diagnosis. And this in terms enabling us to not only compare the results obtained by ourselves on the same galvanometer, with one another, but also with those obtained by other observers on galvanometers of any construction, but graduated in absolute units. An idea of the peculiarities of the absolute galvanometer, contrasted with the galvanoscope, will be obtained by a glance at the diagram. It represents the dial of an instrument divided in its upper half into degrees, in its lower half into milli-

FIG. 21.

amperes. The principle that the angle of deflection does not increase proportionally to the current strength, is illustrated by the fact, that whilst for instance a current of 30 m. a. deflects the needle to about 45°, a current of 150 m. a. is required to deflect it to 70°.

The simplest method of graduating the dial of a galvanoscope into sub-divisions of an ampere, is to place the instrument in the same circuit as a tangent galvanometer of which the reduction factor is known (or a galvanoscope already so graduated) along with a constant cell and a box of resistance coils. By means of the latter the current is modified so as to produce a deviation of 1, 2, 3, &c., to 20 or more milliamperes as measured on the standard galvanometer. The corresponding deviations of the galvanoscope are indicated on the dial, and their value in milliamperes written down.

In the absence of a standard galvanometer, a large standard cell is taken of known electromotive force. The most convenient for this purpose is the Daniell, in which the sulphate of copper is replaced by nitrate of copper. The electromotive force of such an element is as near as possible 1 volt.[o] Placing the cell in a circuit with our galvanoscope the resistance of which must be known (call it 50 ohms) and a rheostat. The cell being large its internal resistance may be neglected.

We then introduce 950, 450, 200, and 50 ohms in the circuit. The current in each case will be (Ohm's law): $\frac{1}{950 + 50} = \cdot 001$, $\frac{1}{450 + 50} = \cdot 002$, $\frac{1}{200 + 50} = \cdot 004$, $\frac{1}{50 + 50} = \cdot 010$. In other words, 1, 2, 4, 10, milliamperes. The cell through the galvanometer will give $\frac{1}{50} = \cdot 020 = 20$ m.a. The intermediate sub-divisions are obtained in a similar manner.

The operation should be repeated with the current flowing in the opposite direction through the galvanoscope, as owing to irregularities of construction, such instruments give different deviations on either side of the zero with the same current.

The reduction factor of every horizontal galvanometer contains a variable element, viz., the magnetic force at the particular point of the earth's surface, which enters into account in the graduation of galvanoscopes. Hence this graduation is absolutely accurate only for such places where the magnetic force is the same as that of the locality where the graduation has been performed. For medical purposes, however, this source of error would be felt only if a graduated galvanometer were to be used in a place far distant from that where it had been made.

The intensity of the terrrestrial magnetism varies also at the same place, from year to year, but in an amount altogether negli-

---

[o] The usual Daniell's cell with HSO_4 (1 : 4) has an E.M.F. of 1·79 volt; with HSO_4 (1 : 12) of ·978; with NaCl (1 : 4), of 1·06 volt. Either of the latter might be taken, allowance being made for the slight difference between its E.M.F. and the volt, (adding 5 ohms to every 100 in the first instance; subtracting 2 ohms in the second).

gible when approximate measurements only are required. This amount is about ·004 yearly.

To give an idea of the variations to which the deflections of the needle are liable from variations of the earth's magnetism, I subjoin the following table of the approximate magnetic intensities in various European cities :—

In the Year	1870	1875	1880
In Paris	1·94	1·96	1·98
In London	1·78	1·80	1·82
In Leipzig	1.86	1·88	1·90
In Darmstadt	1·91	1·93	1·95
In Edinburgh	1·62	1·64	1·66
In Zurich	2·00	2·02	2·04
In Dublin	1·67	1·69	1·71
In Turin	2·07	2·09	2·11
In Vienna	2·05	2·07	2·09
In Königsberg	1·79	1·81	1·83

The angle produced by a given current on a given galvanometer being inversely proportional to the directive influence of the magnetic force acting on the needle, it is obvious that the indication of a galvanometer graduated in London, for instance, would be excessive when used in Turin or Vienna, (in the proportion of 182 to 211 and 209 respectively), deficient when used in Edinburgh or Dublin, (in the proportion of 182 to 166 and 171 respectively). It will be noticed that the magnetic intensity goes increasing as one goes eastwards and southwards from London. The vertical galvanometer, as we saw, is not subject to this source of variations in its indications.

Why is a graduated galvanometer always useful, often necessary in the medical applications of the galvanic current? That a galvanoscope at least is indispensable, is recognised by all those whose experience entitles to speak with authority on the subject. The enormous variations in the resistance offered by the human body to the current according to the individual and the details of the application make even an approximate estimation of the current strength by the number of cells used, impossible. The electromotive force of the cells themselves, and their internal resistance, differ from battery to battery, and from time to time in the same battery. The deviation of the needle alone can give us any information as to what is going on in the circuit, and allows us to compare the currents and ensure their uniformity when required. The sensory and motor phenomena elicited are but an empirical and variable index of the current strength.

A galvanometer graduated in milliamperes enables us, in addition to this, 1st, to measure and register for further reference and comparison the strength of current used in every instance; 2nd, to compare accurately, and if a rheostat be at hand, to measure the resistances in circuit; 3rd, to measure in the same way the electromotive forces in circuit. Such a galvanometer is therefore a means of education in the hands of him who uses it. Physical science is defined as "the comparison of magnitudes." Anyone who wishes to obtain a fair grasp of any one branch of science must practice measurements. Nowhere is such a practice more imperative than in electrics which deals with the intangible. Electro-therapeutical writings swarm with statements proving the impossibility there is to think correctly on the subject, as long as the chief manifestations of electricity have not become embodied, through the channel of the senses, into concrete realities, into measurable magnitudes. A few days spent in observation with a cell, a rheostat, and a galvanometer, will save many a misunderstanding and make a man a better electrician than as many months' reading about Ohm's law and its consequences.

The graduated galvanometer is in medical practice a measurer of doses. As the scales and weights enable us to administer so many grains of iodide of potassium, so it enables us to administer so many milliamperes of electricity. It is true that the mere strength of the current flowing in this circuit is not the sole index to the quantity of electricity circulating in the organ we wish to influence. The size of the electrodes and their relative position, and the depth and surroundings of the organ ought to be determined before we could measure that quantity. But so with the dose of iodide: the degree of its dilution in the fluids of the body and its rate of elimination in each case should be known before we could obtain numerically comparable data. But we are not so advanced in our therapeutical generalisations as to need such a precision. The question is still open whether 10, 40 or 80 grains of iodide is a fair average dose when certain results are to be obtained. Likewise whether 5, 10 or 20 milliamperes is the proper strength to apply, (for a long or short period and by means of large or small electrodes), for the relief of this or that symptom is an ever recurring question in electrotherapeutics. The answer must obviously be limited to the average human subject, and abnormalities or idiosyncrasies be left to be dealt with according to personal experience: still the value of definite average indications can scarcely be overrated.

Much of our posology is still very much a matter of opinion. The more reason then to bring all the methods into play which, though not eliminating every source of fallacy, assist in bringing to a focus the results of a multifarious experience.

ARRANGEMENT OF ELEMENTS.

A subject of practical importance in itself, and which will serve to illustrate the principles established previously, is the arrangement of cells in order to form batteries. By this, we mean the mode of connecting the cells of which the latter are composed. We may, for instance, connect all the zincs together to the one terminal (negative pole), and all the coppers together to the other terminal (positive pole) of the battery; or again, join the copper of the first cell to the zinc of the second, the copper of the second to the zinc of the third, and so on, then attach the negative terminal of the battery to the zinc of the first cell, the positive to the copper of the last.

ARRANGEMENT OF ELEMENTS. 33

The six cells in fig. 22, are connected according to the first method (in surface). In fig. 23, they are connected according to the second (in series). In the former, the current has to pass only once from zinc to copper, inside the cells; in the latter, it has to do so six times.

FIG. 22.

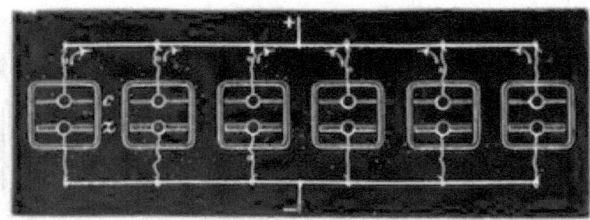

FIG. 23.

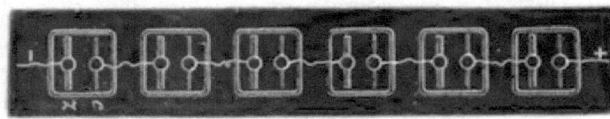

We may also arrange the six cells in two or three series or groups.

FIG. 24. FIG. 25.

 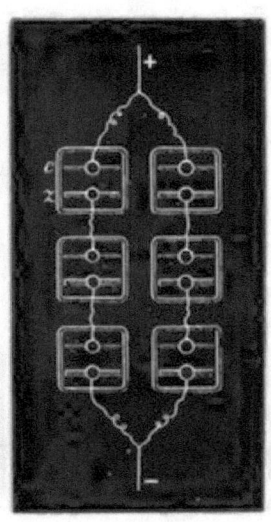

In fig. 24 the cells are arranged in three series of two cells each, or in two groups of three cells; both expressions describe the facts accurately. For it is evident that instead of uniting each zinc of the anterior row to the copper of the corresponding cell in the posterior row, we could unite the three zincs together and the three coppers together (as in fig. 22), and unite the compound zinc to the compound copper by one transverse piece. Likewise in fig. 25 we have two series of three cells each, which might also be called three groups of two cells each and joined as above. In the former instance the current passes twice, in the latter three times, from zinc to copper inside the cells.

Cells united zinc to zinc, and copper to copper, are said to be united in simple circuit or in *surface*; those united zinc to copper, to be united in compound circuit or in *series*.

According to a now exploded theory, the first arrangement is still sometimes called "for quantity," the second "for tension," or "for intensity." The truth is that both are for current strength, and what is called arrangement, for "tension," is nothing more than an arrangement for "quantity" when the external resistance is great. Take for instance, six Daniell's elements. The electromotive force of each is very nearly one volt; and the internal resistance may be taken at 10 ohms. If we join them in series, (fig. 23) we have a total electromotive force of 6 volts; if we join them simply, we have an electromotive force of 1 volt, which is (fig. 22) the same as that of a single cell, the only difference being that the six cells may be considered as forming one large cell, with plates of six times the area of the single one. To use a rough comparison: If we pile up six bricks one foot in all dimensions, we obtain a difference of level of six feet above the ground; if we put them side by side, we have a difference from the ground of one foot only, but an area of six square feet. Suppose now instead of bricks we have pieces of water-pipe of the same size, and arrange them in the same way; in the first case we shall have one pipe six feet long and one foot square; in the second, a pipe one foot long and six feet square. It need not be pointed out that the amount of water these two pipes can transmit and the pressure in them are very different. The superficial analogy between the bricks and pipes and our six Daniells, is sufficient to illustrate what is meant by saying that, when arranged simply, the battery has an electromotive force (difference of level) equal to one, whilst its total surface or conductivity, is six times that of a single cell (that is, its resistance, the reciprocal of conductivity, is six times less), and that when arranged in series, its electromotive force is multiplied by six, whilst its conductivity is six times less, (or its resistance six times greater) than that of a single cell. In the first case the whole electrical current has to pass from zinc to copper, only once, and in doing so encounters six times less resistance than in a single element, because the surface of the metals is six times larger. It is as if one cell six times the size of the original ones were used. In the second it has to do it six times in succession; each time it does so, on the other hand, it receives an increment of electromotive force equal to the difference of potential between copper and zinc, *i.e.* of one volt. (Compare these statements with experiments V and IX.)

In the arrangement shown in fig. 24, the current has to pass twice from zinc to copper, hence the electromotive force of the battery is equal to two volts. It encounters likewise the resistance of the liquid twice—but each time through an area three times greater than that offered by a single cell. It is as if two cells of three times the size of the original ones were connected in series. Likewise, fig. 25 shows an arrangement equivalent to the connecting up three cells in series, each of which was twice the size of the original ones.

Let us now return to the formula for Ohm's law, $C = \frac{E}{R' + R''}$. A battery of 20 Daniells arranged 1st in series, 2nd simply, 3rd in groups, will give us a current through an external resistance R', (the internal resistance, R', being taken at 10 Ohms per cell):

A. Arranged in series: $C = \frac{20 \times 1}{(20 \times 10) + R''}$.

B. Arranged simply: $C = \frac{1}{\left(\frac{10}{20}\right) + R''}$.

C. Arranged, say, in five groups of four cells each, (or in 4 series of 5 cells each).

$$C = \frac{5 \times 1}{\left(\frac{5 \times 10}{4}\right) + R''}.$$

Suppose three cases where R" is respectively a part of the human body to be galvanised; a small portion of animal tissue to be electrolysed; and a piece of platinum wire to be heated for the galvano-cautery.

The resistance of a human arm, with given electrodes, may be taken at 3000 ohms; that of the blood of an aneurism (the two poles being passed in) at about 8 ohms; that of the platinum wire at $\frac{1}{4}$ ohm. We can tabulate the results obtained in these three cases, with the three arrangements above mentioned:

Arrangement.	Arm.	Aneurism.	Wire.
A. Compound or in series.	$\frac{20 \times 1}{20 \times 10 + 3000}$ $C = \cdot 0062$ amp.	$\frac{20 \times 1}{20 \times 10 + 8}$ $C = \cdot 096$ amp.	$\frac{20 \times 1}{20 \times 10 + \frac{1}{4}}$ $C = \cdot 1$ amp.
B. Simple or in surface.	$\frac{1}{\frac{10 + 3000}{20}}$ $C = \cdot 00033$ amp.	$\frac{1}{\frac{10 + 8}{20}}$ $C = \cdot 12$ amp.	$\frac{1}{\frac{10 + \frac{1}{4}}{20}}$ $C = 1$ amp.
C. Mixed, or in 5 groups of 4 each.	$\frac{5 \times 1}{\frac{5 \times 10 + 3000}{4}}$ $C = \cdot 0017$ amp.	$\frac{5 \times 1}{\frac{5 \times 10 + 8}{4}}$ $C = \cdot 25$ amp.	$\frac{5 \times 1}{\frac{5 \times 10 + \frac{1}{4}}{4}}$ $C = \cdot 38$ amp.

The relative strength of the currents obtained in the three cases with each of the three arrangements, shows at once which of the latter is to be preferred, in the particular instance given, to obtain the greatest effect possible from the 20 cells.

We see how it is that in order to heat a wire, we must have a battery, the electromotive force of which need not be high, (one to two volts is enough), but of which the internal resistance must be low; whilst on the other hand in order to galvanise a part of the human body, we must have a comparatively high electromotive force, but may without prejudice use batteries of great internal resistance. There is no cell or battery which may be called "the best" absolutely; everything depends upon the external resistance to be overcome. By making E larger we necessarily increase R': now where R' is large this increase is usually negligible, but not so where R" is small. In the latter case a limit is very soon reached where the advantage obtained by increasing E, is neutralised by the disproportional growth of R' with reference to R".

The best method of arranging a given number of cells so as to produce the greatest possible effect with a given external resistance, is determined by the following calculation: Multiply the number of cells (N) by the quotient of the external resistance R" by the internal R'. The square root of the number obtained indicates the number of the groups into which the cells must be arranged. It will be found that the most favourable arrangement of a given number of cells for giving the strongest current possible with a given external resistance, is the one in which the internal resistance is equal, or as nearly equal as possible, to the external resistance.

Let us apply our formula $x = \sqrt{\dfrac{N \times R''}{R'}}$ to the examples just given. In the cases of the wire and aneurism the results are clear: $\sqrt{\dfrac{20 \times \cdot 5}{10}} = \sqrt{1} = 1$ group; and $\sqrt{\dfrac{20 \times 8}{10}} = \sqrt{16} = 4$ groups But in the case of the arm we obtain:

$$\sqrt{\dfrac{20 \times 3{,}000}{10}} = \sqrt{6{,}000} = 77 \text{ (nearly)}.$$

This result seems at first sight anomalous. But what it means is this, that in order to obtain the strongest possible current from our 20 cells in a circuit, with an external resistance of 3,000 ohms, we should divide them into 77 smaller cells; *i.e.*, that the current obtained by about 80 cells, each a quarter of the original cells, and consequently with four times the original internal resistance, is the strongest obtainable from any arrangement (including subdivision) of our 20 cells. Here also the internal resistance is found to be made as nearly as possible equal to the external, $77 \times 4 \times 10 = 3080$.

Thus we reach the general rule, that in order to obtain the strongest current possible with a given number of elements through various external resistances, we should be able to alter their arrangement with every change of external resistance, so as to make in each case the internal resistance equal to the external.º This principle finds a direct application in galvano-caustics, where provision should be made that the internal resistance of the battery may, by various combinations of its elements, always be made equal to that of the platinum wire used.†

In batteries used for medical purposes, the elements are always arranged in series because the aggregate internal resistance of the cells never come into consideration on account of the large resistance of the body, and of the very moderate strength of the currents applicable to the living organism. The appended

FIG. 26.

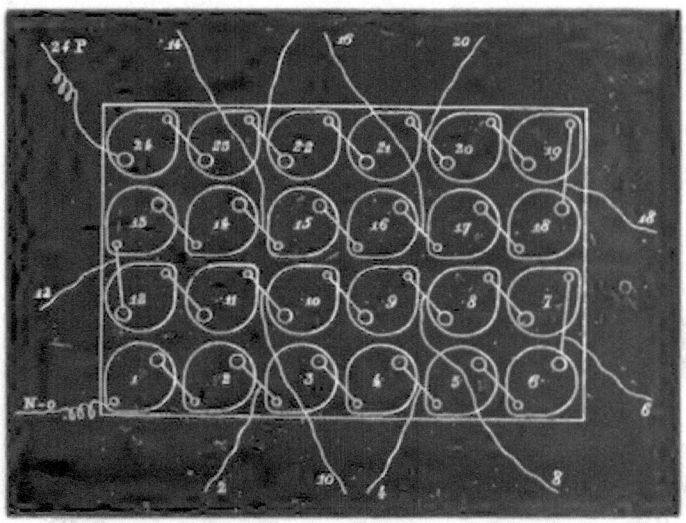

diagram shows a battery of 24 cells connected in series. The wire N-o forms the negative pole of the battery, and is attached to the 1st zinc; the wire 24 P, forms the positive pole, and is attached to the last (24th) copper or carbon. Cross wires are seen to connect

---

* Some writers have committed the error of taking this statement to mean that *under any circumstances*; the best battery is the one in which the internal resistance is equal to the external resistance, to be overcome. This is pure nonsense; the statement concerning the advantage of making the internal resistance equal to the external, applies only to cases where a given number of given cells has to be used for a certain purpose. Whenever we are free to choose, we shall evidently prefer the cells in which, cæteris paribus, resistance is smallest.

† In arranging a battery for the cautery, it must not be forgotten that the resistance of a wire increases by about $\frac{1}{273}$ for every degree centigrade its temperature is raised. We must then make our calculation, not from the resistance of the wire when cold, but when raised to the temperature required. The resistance of a piece of platinum at low red heat (550º) is about twice, at white heat (1300º) six times greater than at 0º.

carbon to zinc of successive cells. The wires marked 2, 4, 6...20, 22, 24, are soldered to the cross wires. In the usual batteries they are connected with the studs of the "collector" which will be described further on. By their means it is obvious that one can make use of a greater or less number of cells (taken by twos) according to necessity. Thus we may attach our electrode (or sponge-handle) to 0 and to the 2nd, 4th or any other wire, in which case the cells in circuit (*i.e.* in use) will be only those included between these wires. We may in fact bring into play any even number of consecutive cells in the battery by fixing our electrodes to two wires so as to include the former in circuit. Thus if we fix them to wires 8 and 14 we bring into action cells 9 to 14 inclusive, and so on. The actual methods by which the cells may be rapidly brought into play, will be found explained under the head of "collectors."

CONSTANCY OF CURRENTS.

The constancy of the current, that is the evenness of its flow during a certain length of time, is disturbed by:—

1. Changes occurring in the cells.
2. Changes occuring in the external circuit.

These changes cannot be altogether avoided, but they can be neutralized to a great degree.

Changes within the cells depend upon the chemical action which is the very condition of the existence of a current. They consist in (*a*) diminution of the electromotive force; (*b*) increase of the internal resistance; and (*c*) the occurrence of polarisation.

Polarisation is a phenomenon due to the evolution of hydrogen and oxygen in the cell from the electrolysis of the water. The hydrogen collects on the "platinode" or negative metal, the oxygen on the "zincode," or positive metal, and a counter current is generated from the hydrogen to the oxygen which may acquire such a strength as to neutralize completely the original current. It is usual, therefore, to interpose between the two metals a "depolarising" substance, *i.e.*, a substance that will combine with the hydrogen and prevent its accumulation on the negative metal. In the Daniell this substance is sulphate of copper, in the Leclanché pyrolusite, in the Grove and Bunsen nitric acid, in the Grenet bichromate of potassium. All these substances are decomposed by, and combine with the hydrogen, and so prevent the occurrence of polarisation. This they do more or less completely, and for a longer or shorter time; and as they are destroyed in the process, the depolarising power of the cell gradually diminishes with use. A fresh supply of the substance has therefore to be introduced from time to time into the cell.

Experiment XV.—Take a piece of zinc and of platinum wire and immerse them in acidulated water. Connect them with a galvanometer. The deflection speedily diminishes. Now drop some nitric acid in the neighbourhood of the platinum: the deflection does not diminish so rapidly, and at the same time fumes of nitrous acid are given off. The hydrogen reduces the nitric acid, and does not adhere to the platinum.

Simultaneously with these phenomena, alterations take place in the electromotive force and internal resistance. Both of these, as we know, remain constant only as long as the constituents of the cell are uniform in composition. But as the zinc is dissolved and fresh compounds are set free within the cell, these constituents are altered, electromotive force falls, internal resistance rises: the cell becomes weaker. The rate of these changes is proportional to the activity of the cell, that is to the strength of the current used and the time during which it flows, *i.e.*, to the quantity of electricity evolved. If therefore, a large and a small cell are put to the same service the larger will last longer, because there is a larger supply of materials within it.

For the same reason also polarisation is more efficiently neutralised in larger cells, which consequently are more constant. Medical batteries, unfortunately, require a large number of cells and must generally be made portable; hence medical elements must be small, and consequently less constant and durable.

The changes occurring in the circuit during the passage of a current, when a given portion of the body is included between two given electrodes are:—

Polarisation of the electrodes and tissues.

Permeation of the epidermis with moisture, and hyperæmia of the subjacent structures.

The former set of changes tend to weaken the current, just as polarisation within the cell does it, but not to a very appreciable degree in ordinary medical galvanisation.

The latter by diminishing the resistance of the integuments, materially augments the strength of the current at least during the first few minutes of the application.

The importance of these sources of inconstancy can hardly be over-rated, especially in electrodiagnosis. They cannot be avoided, but can be met in two ways:—First, by correction; that is by the combined use of the galvanometer and current regulator, the number of cells in circuit being constantly adjusted so as to keep the needle steady. This explains itself, and is the system most readily adapted to medical batteries. Second, by elimination. This method consists in using a very large number of cells and intercalating in the circuit a very large additional resistance; any oscillations in the resistances of the body, electrodes, due to the causes above described, become insignificant—nay the whole resistance of the body, etc., may be neglected in comparison to that additional resistance. Thus, if I wish to pass a current of a certain strength, say of 5 milliamperes, through a part of the body of which the resistance may be taken at 3000 ohms, I must use 15 Daniells, (taken at 1 volt each), for $\frac{15}{3000} = \cdot 005$ current strength. The same current strength will be obtained from 250 Daniells through a rheostat resistance of 50,000 ohms: $\frac{250}{50000} = \cdot 005$. It is clear that the throwing into this circuit of the 3000 ohms of the body will only produce a difference of about $\frac{1}{16}$ only in the total current strength. Much more, therefore, will any oscillations within the resistance of the body fall out of calculation, and the utmost constancy be secured.

For purposes of diagnosis and physiological investigation especially, the plan of using a larger number of elements with an additional resistance is invaluable, as it secures a great accuracy in the results, and saves much time by making the observer independent of the minute attention to the details of the operation otherwise indispensable. The large numbers just mentioned are typical; but the principle remains the same when smaller batteries and smaller additional resistances are used. The advantage derived from the method of elimination proportional to the ratio between the additional resistance and that offered by the body.[*]

As previously mentioned, currents impelled by a high electromotive force through a high resistance are called "currents of high tension" in the language of the old fashioned text-books. This is a convenient expression, and may be preserved if it be carefully divested from the exploded theories with which it is associated.

DERIVED CURRENTS.

IF two or more paths are opened to the current, as when, for instance, the poles of a battery are connected by two or more wires, the current divides itself among the wires in such a way that the strength of each branch current is inversely proportional to the resistance of the wire. The same thing occurs when a cistern is provided with several exit-pipes, the amount of water flowing through each is directly proportional to the diameter of the pipes, or, if we suppose the pipes of equal diameter and length but unequally packed with pieces of sponge, inversely proportional to the resistance offered to its flow in the several pipes.

Let B represent the battery, and let the rheophore bifurcated at

FIG. 27.

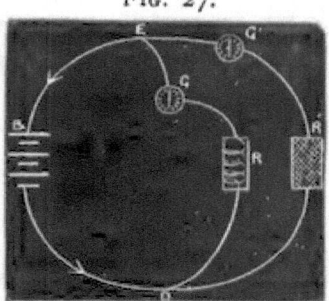

[*] The use of this method in electrotherapeutics was first advocated by me in a paper in the *Medical Times*, Sept., 1877. It has since been applied to the investigation by Tschiriew and myself of the excitability of cutaneous nerves (*Brain*, 1879) and to that by Waller and myself of the Electrotonus of man. (*Philosophical Transactions of the Royal Society*, 1881.

D convey the current to a rheostat R and to a part of the human body R'. A galvanometer is included in each of the derived currents, G, G. Suppose the resistance R' to be 2000 ohms. First make the resistance in R 2000 ohms, the galvanometers will indicate that the currents passing through R and R' are equal. Next increase the resistance in R to 3000, 4000, &c.; the deflections show that the currents in R and R' become as 2 : 3, 2 : 4, etc. Finally, diminish the resistance in R to 1000, 500, etc., and the currents assume the ratio in R and R' of 2 : 1, 2 : ·5, etc.

By inserting a third galvanometer into the portion of the circuit E B D, where the current is undivided it is seen that the sum of the currents in R and R' is equal to the current in E B D.

DENSITY OF CURRENT.

By the density of a current is meant the proportion between the Quantity of electricity, or water, it conveys, and the Diameter of the conductor through which it flows, $\frac{Q}{D}$; the larger D the less the density, and inversely. A current of 1 ampere is ten times denser when it flows through a wire of 1 millimetre in diameter, than when it does through one of 1 centimetre; ten times more electricity passes through every unit of the sectional area of the first wire than through every unit of the second. Density of current therefore may be described as the quantity of electricity flowing through a unit of the sectional area of the conductor in the unit of time.

A clear conception of the meaning of the term is of vital importance to anyone who wishes to understand the behaviour of currents applied to the human body. At the risk of appearing tedious, I shall therefore present the notion of density over again, under different aspects. If we send a current of 1 ampere, (that is conveying one weber in the second) through electrodes 1, 2, 5, 10 square centimetres of sectional area respectively, it is evident that whilst in the first case 1 weber will pass through 1 sq. c.m., only ·5, ·2, ·1 weber will pass in the other electrodes through the same area: in other words the density (*i.e.* the chemical and physiological action) of the current in the part of the body under the electrode diminishes with the size of the latter. If again our current flows through a wire, the diameter of which alters from place to place, we shall say that its density at any part of the circuit, is inversely proportional to the diameter of the conductor, at that part. If the circuit is composed of heterogeneous materials such as metallic wires, a pair of electrodes, and the human body, the density of the current flowing through will likewise be, at any point of the circuit, inversely proportional to the diameter of the conductor at that point; densest therefore in the wire, less dense in the electrodes and in the part of the body immediately under the electrodes, least dense in the portion of the body midway between the electrodes (see figs. in the next paragraphs and in the chapter on electrisation).

The reader may picture to himself the electrical density at any point of a circuit of variable diameter, by representing the strength of a given current flowing through it, by a certain number of lines (as is done in subsequent diagrams). These lines expand in the wider portions of the circuit, owing to the diffusion, and become crowded together in the narrower parts. A crowd issuing through a narrow door and gradually expanding passages and finally reaching the street, like electricity flowing through a circuit of variable diameter, is said to be densest at the narrow exit; it thins out, is diluted in space, as it were, as it reaches the wider outlets.

DIFFUSION OF CURRENT. DISTRIBUTION OF POTENTIALS.

WHEN the rheophores of a galvanic battery are placed in contact with any two points of a conductor of any shape and size, the whole of the conductor is permeated with electrical currents, which diffuse themselves in accordance with the laws of the derived currents, *i.e.*, which are the weaker the longer and more circuitous the course they take if the conductor be homogeneous; or, what comes to the same thing, the greater the resistance they encounter. It is evident that a conductor of any kind can be looked upon as consisting of layers of conducting substance thinning out towards the points of application of the rheophores. Each of these layers offers a passage to the current, which divides itself among them, (exactly as if they were wires) into derived currents of a strength inversely proportional to the resistance of their respective paths, and of a density inversely proportional to the sectional area of these paths. A glance at the diagram (fig. 19) where the lines of current from P to N subdivide the conductor into such imaginary layers will make this evident.

In order that a current should flow from one point to another, there must be a difference of potential between these two points. Hence different parts of a conductor through which a current flows are at different potentials. With a cylindrical homogeneous conductor, such as a wire, the distribution of these potentials is very simple: such a conductor may be conceived as made up of a series of infinitely thin disks, each of these disks transmitting the current from the disk preceding to the disk following it. Every part of such a disk would then be at the same potential, which potential would be intermediate between those of the contiguous disks. In other words, every point of the surface exposed by a transverse section of a wire is at the same potential: that is, such a surface is an equipotential surface.

The case is more complicated when the conductor is not evenly pervaded by a current, such as in the case of the rectangular uniform conductor represented in the diagram. The equipotential surfaces are here represented by the curves 6 6, 5 5, etc., o o,—1—1 etc. for which, however, I do not claim anything like geometrical accuracy. Suppose the battery C Z, (the internal resistance of which is negligible), has 14 volts electromotive force. The potentials at the points P and N are then equal to + 7 and — 7 volts respectively; and the distribution of potentials in volts along P N is as indicated by the numbers, the middle point being at zero, whilst within the conductor itself the potentials are to be found distributed as roughly shown by the curves 6, 5, 4, etc.

FIG. 28.

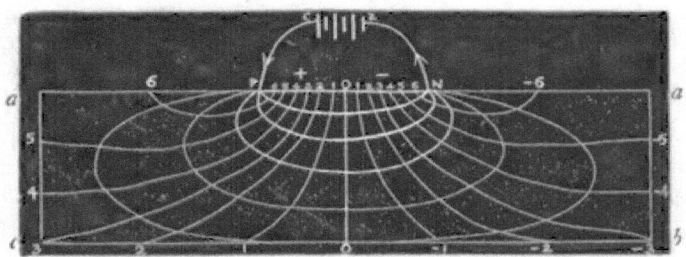

Explanation of Diagram.—In the diagram *a b c d* represents a rectangular conductor with which the rheophores of a battery C Z are brought into contact at the point P and N. The whole conductor is then permeated with currents, the general direction of which is represented by the lines joining P N. By means of an electrometer or galvanometer, it can be shown that the distribution of potentials in the conductor is somewhat as shown by the lines + 6 6, + 5 5, etc., 0 0, — 1 1, — 2 2, etc. If *a b c d* represents a piece of glass covered with a thin metallic film and dusted over with iron filings, when the electrodes are applied at P N, the filings will arrange themselves so as to have their long diameter along the equipotential lines 6, 6, 5, 5, etc., etc. (The lines of current direction P N, in order to give an absolutely accurate representation of the facts, should cut the equipotential lines wherever they meet, *at right angles*).

THE HUMAN BODY AS A CONDUCTOR.

ANY one who wishes to apply the current intelligently to the human organism, should spare no trouble to master completely the facts presented in this paragraph. This cannot be done by mere reading, but only by the sedulous practice of the experiments mentioned here and in the chapter on electrophysiology, in the light of the general principles of electrophysics laid down previously, (Ohm's law).

EXPERIMENT XVI.—Take a battery with galvanoscope in circuit, and choosing a convenient electromotive force, (ten Leclanché cells answer the purpose very well), note the deflection of the needle obtained when two medium sized electrodes, moistened with cold water, are applied to two given spots on one side of the body, 1st, at the moment of application; 2nd, after the electrodes have been kept in situ for five minutes. The deflection in the latter case is greater, (which means that the current is stronger) because the resistance of the epidermis has been diminished by the moisture of the electrodes. Repeat the experiment on the symmetrical points of the other half of the body, with the electrodes soaked with hot salt and water. The deflection is greater at first, and soon reaches a greater maximum than before, because the penetration of the fluid into the skin is more complete and rapid. Note also the fact that the current is stronger when the electrodes are applied firmly to, than when they are made to touch lighly, the surface of the body.

EXPERIMENT XVII.—Take a pair of each of the different sizes of electrodes (fine, small, medium, large, very large), and apply them successively to the same two spots with the same electromotive force (10 cells) in circuit. The galvanometric deflection increases with the size of the electrodes.—Repeat the experiment taking the electrodes in the opposite order, so as to eliminate the fallacy arising from the increased moistening of the epidermis; and take care that, throughout, the electrodes be uniformly soaked.—The skin forming the chief resistance of the circuit, every increase in the area it offers to the entrance of the current, is accompanied with a considerable increase in the total current strength.

EXPERIMENT XVIII.—Take two medium sized electrodes and with 10 cells in circuit observe the deflections you obtain when they are applied to a number of symmetrical points on the face, neck, trunk, and extremities. You will find that the deflections vary greatly.—The experiment must be made carefully; the electrodes are to be well moistened each time they are applied; and each time left in situ for the same number of seconds.—That these differences are due chiefly to the varying skin-resistance, can be demonstrated by repeating the experiment with needles thrust into the subjacent tissues (as is done in the treatment of rheumatism, paralysis, etc., by acupuncture), and noting the deflections given with one or two cells in circuit.

EXPERIMENT XIX.—Apply a large plate electrode to the nape of the neck, and when the skin underneath it has been thoroughly soaked, apply the other electrode (small or medium size) to a series of points along the trunk and limbs, taking the same precautions as in the former experiment. Note the deflections obtained in each case: you will find that they vary, not according to the distance between the electrodes, but according to the point touched with the moveable one. The substitution of a needle for the latter, or the occurrence of superficial ulcers or excoriations will serve to show that the variations observed, depend upon different skin resistances at different points of the body.

It is of great service in the electrisation of the human body for diagnostic or therapeutic purposes, to be familiar with the relative resistance encountered by the current, at different points of its surface. Such a familiarity will speedily be obtained by systematically observing the deflection of the needle, under varying conditions of experimentation. Though a galvanoscope is sufficient to demonstrate the existence of great differences between the body resistances, the observer will be greatly assisted in the realisation of the facts by using a galvanometer divided into units of current strength, which allows him to compare these resistances with one another, and a rheostat which enables him to measure them.

Looked upon as a conductor of electricity, the human body may be compared to a vessel bounded with poorly conducting material (the skin) unequally packed with non-conducting solid particles, the interstices being filled up with a saline fluid of fair conductive power. The parts most densely packed with solid particles are represented by the bones; those where liquids predominate, by the muscles. Between the two are found the nerves, viscera etc. The contents of the blood vessels are the best conducting media in the body. In short the list of the various tissues arranged according to their watery constituents may be considered as classifying them according to their electrical conductivity. The diffusion of electrical currents in such a conductor is evidently very irregular, and impossible to represent accurately either by word or drawing. It has no specific resistance, nay it is impossible to define accurately the specific resistance of any one of its component parts whether nerve, bone, or muscle, etc. The unequal distribution of blood among them, their complex structure (not to speak of their unequal internal polarisation) makes any statement concerning their resistance subject to variations within wide limits.

The chief resistance of the skin is located in the epidermis, which when dry may be considered as an insulator. The epidermis however varies in thickness at different points of the body, and is not continuous, being perforated more or less thickly with the outlets of sweat-glands and follicles. Its resistance diminishes in proportion to the degree of its moisture, (artificial soaking, perspiration). The increased vascularity of the underlying cutis has probably some influence in reducing the resistance of the integuments.

The resistance of the skin varies accordingly, to a prodigious extent, not only between different individuals and between different parts of the body of the same individual, but at the same points of the same individual under different conditions.

Some minor points have to be taken into consideration with reference to the altered resistance of the epidermis, through its permeation with water. Hot salt water reduces this resistance more than pure cold water. The cataphorical property of the current, and perhaps also its influence on the cutaneous vessels have the effect of reducing it still further than mere soaking. (The effect of a current on the deeper tissues—polarisation—which is manifested on reversing it after it has flowed for a time, will be best considered along with its physiological effects).

A most important conclusion to be drawn from the fact that the skin, even under the most favourable circumstances, offers the chief resistance to the current, is that the size of the electrodes, that is to say, the extent of the body-surface offered to the current as spots of inlet and outlet, to a great extent determine the strength of the current obtainable from a given number of cells, through the body. The resistance of a conductor, we know, varies inversely with its diameter: hence by doubling, trebling, etc., the size of the electrodes, we shall diminish the resistance of the skin 2, 3, etc. times. Now since this resistance forms the greater part of the total resistance in the circuit, it is obvious that its diminution will be of great influence on the strength of the current. If we assume the resistance of the skin, to be 100 times greater than that (R) of the rest of the circuit (battery, electrodes, internal tissues of body), the current strength, according to Ohm's law, will be given by the equation $\dfrac{E}{100 R + R}$ where E is the electromotive force in circuit.

It is obvious that by diminishing the resistance of the skin by one half, we shall increase the current in the proportion of 1 to 2 nearly for $\dfrac{E}{\frac{100 R + R}{2}} : \dfrac{E}{100 R + R} = 51 : 101$. These numbers are not meant to represent the facts, but merely to illustrate the principle involved in the problem.

The above discussion may be conveniently summed up in the following statement.

"The resistance of the human body" when put forth as a general statement is a meaningless expression. By it, is to be understood chiefly the resistance of the two particular portions of the skin in contact with the electrodes under the conditions of the particular application; the resistance of the intermediate portion of muscle or other tissues (except bone) being almost negligible in comparison to the former, partly owing to the better conductivity of these tissues, partly to the large sectional surface they offer to the flow of the current. The resistances of the body will vary not so much according to the distance of the electrodes from one another, as according to their diameter, the degree of moisture of the epidermis, its thickness at the point of application, the number of sweat ducts offered to the current, and the state of the cutaneous system, and to a certain degree, with the strength and length of the electrical application. Pressure on the electrodes diminishes the resistance.

These considerations help to explain the prodigious discrepancies between the numbers ascribed by various observers to "the resistance of the body." The numbers given in every instance can apply only to the particular parts of the body of the particular individuals, tested under particular circumstances; when therefore we read or say that 2,000 or 3,000 ohms may be assumed to represent the average resistance of the parts of the body when galvanised with medium electrodes according to the polar method adopted in ordinary medical applications, no practical importance, except from the most general point of view, can be attached to statements of this kind, which are used for the purpose of argument and illustration only. Variations ranging between 1,000 and 10,000 ohms, are by no means rare under those very conditions.

We now come to the diffusion of the current in the body. Wherever on the surface the two electrodes are placed, the whole of it from the soles of the feet to the scalp and tips of the fingers, has its potential altered. There occurs in it the same phenomenon as that pictured in the diagram illustrating the distribution of potentials, in a homogeneous rectangular conductor. Equipotential lines of very irregular outline might be demonstrated in it, at least on the dead subject, extending throughout to its remotest parts, and currents permeate its whole substance flowing at right angles to the equipotential lines. The existence of these currents can be made manifest, by inserting two needles connected with a multiplying galvanometer, into the parts remotest from the electrodes, (*e.g.* in the foot when the latter are placed on the arm).

It is of the highest importance to realise all the consequences of current diffusion in electrotherapeutics. We shall return to this subject more in detail, when discussing the conditions of nerve-excitation on the living body and electrisation of diseased organs, (see diagrams in subsequent chapters).

From the fact that the specific resistances of various tissues and organs are widely different, and from the laws of diffusion just set forth, we may put down as a practical axiom that comparatively powerful currents are required to influence, *i.e.* reach with sufficient density, those tissues which being poor conductors themselves are surrounded with better conducting material, such as nerves deeply embedded among muscles.

FIG. 29.

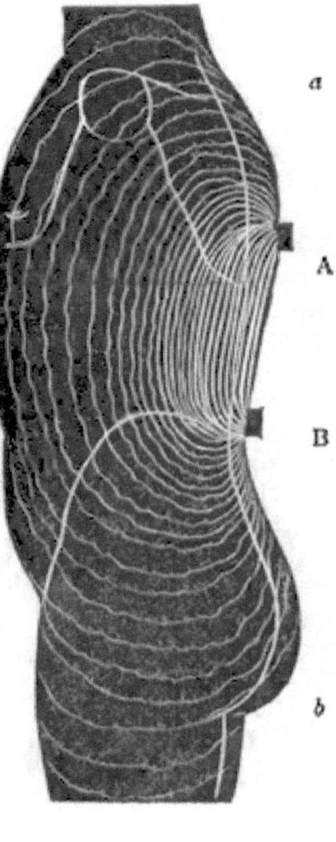

The diagram fig. 29 is intended to illustrate the fact that the current diffuses itself throughout the body whatever the points of application of the electrodes may be. Absolute localization is impossible. If A and B are the positive and negative poles respectively, the bulk of the current flows in more or less curved lines from A to B through the tissues enclosed between them. Externally to this interpolar region, the direction of the derived currents proceeding from each electrode is opposed to that of the main current in the region, A to *a*, and B to *b*. We assume here that the body is made up of a uniformly conducting substance. The thick lines denote the region of greatest current density.

It is of the utmost importance to bear in mind the diffusion of currents in electro-diagnosis where it is the source of numerous fallacies (contraction of distant muscles, etc.). The occurrence of cerebral symptoms (giddiness, flashes of light, the "galvanic taste") when one of the electrodes is applied to the upper part of the back, is due likewise to the diffusion of current upwards.

FIG. 30.

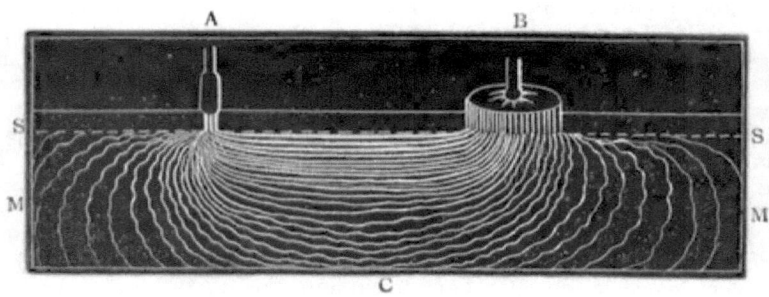

The diagram fig. 30 is intended to illustrate the following facts:—Two electrodes, A B of different sizes being applied to the skin, S, overlying muscular or other tissue, M, the current exerts the most powerful influence where it is the densest, that is under the smaller electrode. It diffuses itself over an indefinite area in the manner I have attempted to depict, but which the engraver has not very faithfully rendered. The shorter the distance between A and B, the greater will be the proportion of the current flowing through the tissues immediately between them; the longer the distance, the more it will diffuse itself through the whole of M; for as A recedes from B, it is obvious that the difference between the length of and the resistance offered by, the direct path A to B, and the length and resistance of the more circuitous path, A C B, diminishes proportionally.* Hence the practical rule that the nearer the electrodes, the denser the current (the more powerful the effect) in the tissues immediately between them.—Through the skin the current chooses chiefly the points of least resistance, *viz.*, the sweat ducts.

MECHANICAL EFFECTS OF THE CURRENT.

WHEN a current is sent through two vessels filled with a liquid and separated by a porous substance, such as an animal membrane, the level of the fluid sinks in the vessel in which the anode is placed, and rises in that containing the kathode. In other words the current produces a transference of the liquid from the positive into the negative vessel. This phenomenon is known as osmosis. It naturally occurs between two dissimilar liquids; for instance between water and a solution of gum, the direction of the osmotic currents being from the lighter to the denser liquid; but the electrical current, when opposed to this natural direction, is sufficient not only to neutralise the phenomenon, but to reverse the order of its occurrence.

The amount of electrical osmosis grows with the resistance of the fluid acted upon, and is thus inversely proportional to the amount of electrolytical action. Thus it is very marked with pure water, and diminishes with the conductivity imparted to the liquid by salts in solution, etc.

* Though the amount of internal tissues, included between the electrodes, does not materially interfere with the strength of the current (which is regulated by the epidermic resistance), it does govern the distribution of the current, or ts density, in the portion of the body traversed.

When a current is sent through part of the body, we may readily suppose that such a transfer of liquids from cell to cell occurs on its path. Though it is difficult to estimate, even approximatively, the therapeutical importance of osmotic phenomena, yet they must be borne in mind when we try to account for the influence of the galvanic current on tissue-change, and for its notable power of promoting resorption in certain cases of effusion.

PHYSICAL EFFECTS.

The current has certain characteristic physical effects, such as the generation of light and heat, the induction of electromotive force in neighbouring conductors, and the magnetisation of iron.

The heating effects of the current are used in galvanocaustics where a platinum wire of appropriate length and size is brought to incandescence by its means.

Incandescent wires are also used for the purpose of giving light (as for instance in Trouvé's polyscope) for the examination of the throat and other internal organs. In the now popular Edison lamp a filament of charcoal, enclosed in a glass bulb devoid of air, replaces the platinum wire.

Some authors speak of the "thermal" effects of electricity in the body, apart from the warmth produced by increased circulation. They are insignificant; medical currents are so weak, and the bulk of the body as a conductor is so great, that *a priori* it is absolutely impossible that a sufficient quantity of electricity should be converted into heat so as to raise perceptibly its temperature even at the points of application of electrodes of ordinary size.

Some authors, again, have laid much stress upon the existence of induced currents in the nerves set up by every make and break of the galvanic current. The existence of such currents is doubtless possible; but it is difficult to estimate their importance with reference to the physiological effects of electricity. Dr. Radcliffe goes so far as to attribute to such induced or extra-currents, the causation of contractions at the make and break of the galvanic current. But he falls into the curious error, fatal to his hypothesis, of assuming that it is the break extra-current that is opposed to the battery current, instead of the make. That the primitive galvanic current can call forth contractions independently of such induced currents accompanying its make and break, is abundantly shown by the fact that muscle deprived of its nerves reacts to galvanic and not to faradic stimuli.

CHEMICAL EFFECTS.

When the extremities of the rheophores connected with a galvanic battery are plunged in water, a more or less abundant evolution of gas is noticed; oxygen is given off at the positive pole or anode, hydrogen at the negative or kathode. This phenomenon is called electrolysis, or decomposition by electricity. The water is split up into its constituent elements. The same thing occurs when the solution of a salt is subjected to the influence of the current. Thus, if iodide of potassium is electrolysed, free iodine appears at the positive pole, and potash at the negative.

Similarly the chlorides, phosphates, etc., of the body, undergo decomposition, the acid radical being set free at the positive, the alkali at the negative pole. If then we expose a piece of animal tissue, or albumen, to the action of the current the following phenomena will be observed.

1. The water it contains is decomposed; oxygen is set free at the anode, and an abundant froth, due to the more voluminous evolution of hydrogen is noticed at the kathode.

2. The salts split into acids at the anode and alkalies at the kathode.

3. The former along with the nascent oxygen oxidise the neigbouring tissues; the latter exert their peculiar caustic action. The tissues attacked form in the one case a hard dry eschar, in the other a soft moist mass, frothy by the admixture of hydrogen bubbles.

When a sufficiently powerful, or sufficiently protracted current is applied to the human body, similar changes occur in the skin, at the point of contact of the electrodes. Under the anode an eschar is produced with an acid reaction; at the kathode a vesicle, filled with an alkaline liquid. If the action is prolonged beyond this point, more extensive ulcerations take place, having the general characteristics of those produced by acids and caustic alkalies respectively.

The use of "impolarisable electrodes" consists in preventing the accumulation of the products of electrolysis at the points of their application. Hence the dangers of producing eschars, and the accompanying pain, are obviated. They are universally employed in physiological experiments; but in medical practice can be so only in exceptional cases owing to the complicated manipulations their preparation necessitates. An impolarisable electrode consists of a glass tube plugged at one end with moist clay, and fitted at the other with a cork through which passes a rod of amalgamated zinc. The tube is filled with a saturated solution of sulphate of zinc. The zinc is connected with one of the poles of the battery, and the clay applied to the skin or nerve.

Intimately connected with the fact of electrolysis is the existence of *currents of polarisation* in the organism at the moment of breaking of the galvanic current. These currents are opposed in direction to the original current, and can easily be demonstrated by placing the hands in two basins connected with the poles of a battery, and after a few minutes electrisation transferring them (after wiping) into two other vessels connected with a galvanometer. The needle will indicate the existence of a current flowing in a direction opposite to that of the primitive current. The same phenomenon is observed in experiments upon pieces of animal tissue, and is obviously analogous to the "polarisation" of inorganic substances.

CELLS AND BATTERIES.

Any galvanic element capable of furnishing a *constant* current of the greatest strength and longest duration required in applications to the human body is thereby qualified for medical uses. Constancy is the only essential quality of a medical battery.

Any claim set up for this or that form of cell on the grounds of its yielding a current possessing specific healing powers of its own, must at once be put down to self-delusion, or to a desire to delude others, on the part of he who puts it forth.

But if all cells of sufficient constancy (and this constancy is a characteristic feature of all our modern forms of galvanic elements) are eligible from a theoretical point of view, the case is very different when we look at it from a practical stand-point. Before an element is fitted to be used in the construction of a medical battery, it has to fulfil several conditions which limit the choice to a very few forms.

A medical battery must usually be portable. Three factors make up portability: Smallness of size, lightness, absence of liquids easily spilled. Now many elements such as the Daniell, the Grove, and other cells cannot be made small and are excluded from our list on this score. The Grove again, the Stöhrer, and generally all elements with acid, are exposed to objection on the ground of their liquid contents being easily spilt when their equilibrium is disturbed. The weight of a 30 celled Leclanché or Stöhrer battery is scarcely less than 30 pounds, including the necessary accessories. Now this is a great weight for a battery which has to be carried about freely. The only element which yields really portable batteries is the Gaiffe (chloride of silver), but it is expensive and to be recommended only to those in the habit of using their battery every day. Much remains to be done in improving medical batteries with reference to their size and weight.

A medical battery must also give the minimum amount of trouble as regards the manipulations requisite to keep it in working order. It ought not to be subject to wear when not actually in use; and should be easy to take to pieces when it requires recharging and cleaning—a process which should be rarely wanted. All cells with powerful chemical action require daily or at least frequent re- and charging are thus excluded from use in medical batteries. The dilute acid cells (Stöhrer's,) with a mechanism to remove the zincs, etc., from the acid during rest, are as a rule easily taken to pieces and put together again, as well as recharged. The Leclanché on the other hand is usually sealed up and cannot be recharged except by the instrument maker; on the other hand, this process is not frequently needed, but only once in a year or two, according to the quality of the materials used in their manufacture, and the uses to which the battery have been put. Notwithstanding the great improvements effected in medico-electrical apparatus during the last few years, the existing batteries can scarcely be said to offer more than a choice of evils. In the absence of the ideal battery, combining the several requisites enumerated above we must be content with adopting the one which offers the least disadvantages, and I think on the whole the Leclanché to be described presently is the one to be preferred, and which adapts itself best to most purposes, unless the new chloride of silver element, now under trial, is found to realise the hopes it has raised.

Batteries are of two kinds; 1st, those used for diagnosis as well as for treatment; 2nd, those used for treatment only.

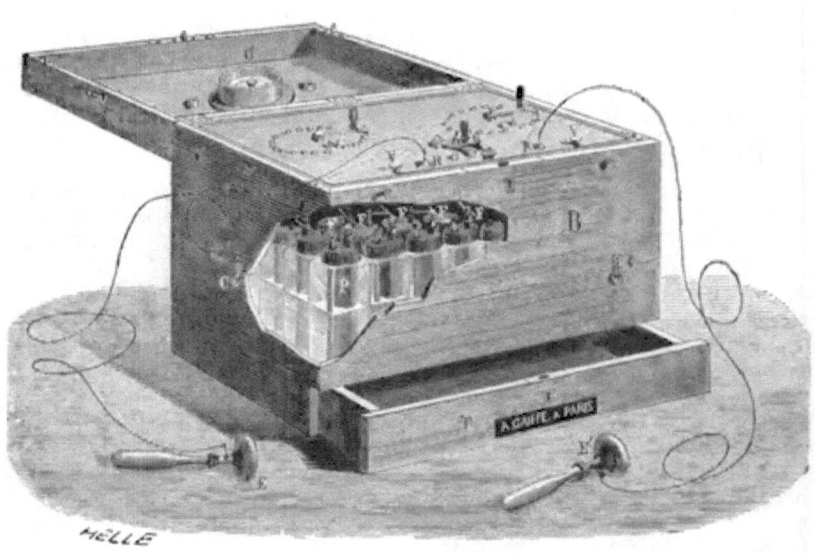

Complete medical battery. The element board carries the double collector, commutator and interruptor described elsewhere. V, V, are the bolts for securing the board, which is hinged so as to allow a ready inspection of the elements.

The lid carries a galvanometer G.

The case is separable into two halves for facilitating the repairing or refilling of the cells when necessary. The hooks, c, c, secure the two halves together. A portion of the case is removed showing the arrangement of the cells P, and the connecting wires F. A drawer is provided for containing the rheophores, electrodes, etc.

A battery (fig. 19) of the first kind should consist of forty cells at least. It should be provided with a dial collector, a commutator and a galvanometer.

A battery (fig. 32) the second kind need not usually have more than thirty cells, and should be provided with some means of taking the cells by twos or threes, and be fitted with a galvanoscope.

When portability is not absolutely wanted, as in the case of batteries for the consulting room or for the electrical room of a hospital, the best plan is to have 50 pint or quart size Leclanchés disposed in single rows on shelves, and a table placed in front, with the

Fig. 32.

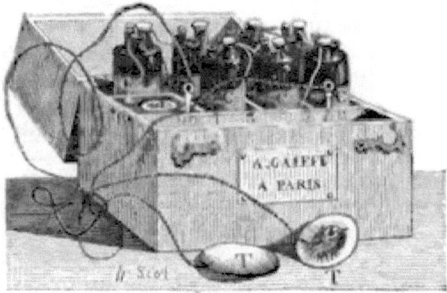

Simple medical battery of 20 cells, with pin and hole collector, and galvanocope, tin plates covered with wash leather.

Fig. 33.

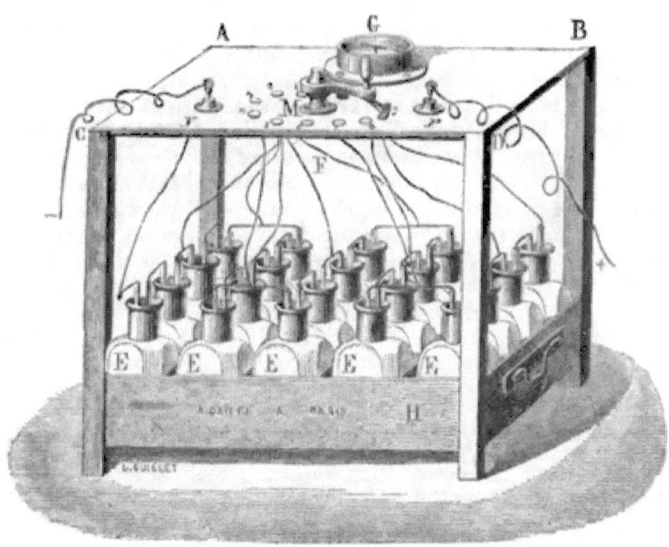

Battery of 20 cells E, E, &c, showing the connections. The element-board carries a switch collector M, a galvanometer G, and the binding screws. The case H is an open framework. This kind of battery is suited for hospitals, or consulting rooms, when a small number of elements only is necessary. The crowding together of a large battery in superposed drawers is extremely inconvenient, and the arrangement described in the text, the only one to be recommended when there are more than 20 or 25 cells.

collector, galvanometer and other accessories fixed on the top of it; along with a sledge faradic apparatus, and the commutator-alternator to be described further on.

Though large cells have no innate advantage for medical purposes over small cells, they are to be preferred as being cheaper in the long run.° They contain a larger provision of the current-generating and depolarising materials, and are less liable to become the seat of secondary chemical changes, when not in use. They need not be sealed down, and may be got at, examined and replenished by almost any intelligent person who has had some experience in electric telegraphy, or by the owner himself who is spared the necessity of summoning the original maker or sending the battery to him.

A few remarks on the chief causes of failure in batteries may conveniently be added here. First, if the failure is gradual, it will probably be found that it is due to the deposition of salts in, or to exhaustion of the cells. In this case, the remedy is obvious: the latter must be cleaned and replenished. This is not the place for entering into the details of the operation, which vary according to the kind of battery used. Instructions on this point are usually given by makers to the buyers of their instruments, when the operation is within the competence of the latter.

Second, the failure is more or less sudden. The source of the mischief may be, 1, in the cells, 2, in the connections between the individual cells, 3, in the connections between the cells and the element board, 4, in the rheophores and electrodes. In order to find out which of these parts is at fault, a methodical investigation is necessary, and the best is to proceed by elimination. The electrodes applied to the skin do not yield any current: disconnect them and find out whether this occurs only at one part of the battery, *i.e.* at one stud of the collector; if so, the cause will be found in the wire connecting the corresponding cell to the element board. If the battery works up to a certain number of cells only, there is a break in the cells, or in the wires connecting them together, just beyond that point. (Compare fig. 26, and description of collectors). If no current, whatever, can be obtained, the electrodes and rheophores should be changed; when if the current is still in abeyance, and

---

° It is perhaps not altogether unnecessary to refute explicitly here the old fallacy that "big cells, having large surfaces of metal, give a greater quantity of electricity than small cells, in which the chemical action cannot be so vigorous." Now we have seen in Experiment I, that in a properly constructed cell no chemical action takes place when the current is not flowing—and the amount of chemical change is, by the law of conservation of energy, proportional to the quantity of force liberated; that is to say to the strength of the current. Now by Ohm's law this strength is governed by the resistance in circuit, the electromotive force remaining the same. Therefore when the external resistance is large in proportion to the internal resistance, the current given by a given element, is practically the same whatever the size of the cell. Thus a zinc-carbon cell with plates one square mile in surface, will give much the same current through the body, as one the size of a thimble. The resistance of the former may be taken to be zero; of the latter to be 20 ohms. Now the electromotive force of each being 2 volts nearly, it is obvious that $\frac{2}{3000} : \frac{2}{3020} = 150 : 151$. (Chap. VI, p. 36).

the whole number of cells being in circuit, the fault will be finally localised in the first wire, (marked O in fig. 26), or the first cells. Having thus localised the source of mischief, it must be remedied accordingly. Electrodes must be cleaned or mended; rheophores must be re-adjusted; faulty connections rectified; exhausted cells replenished.

The rheophores are often at fault, usually at the points of attachment to the electrodes. With ordinary telegraph wire this can hardly escape immediate discovery; but in the old fashioned silk covered conductors, the metallic threads may be completely broken, and yet nothing appear externally.

Defects in the element-board and its connections are not always so readily discovered and rectified; the less so, the more complicated their structure. A certain amount of familiarity with the anatomy of his battery ought to be acquired by its owner, so that he should not be entirely dependent upon the instrument maker for rectifying every little hitch that is sure to occur at some time or other.

This remark applies with equal force to the accidents to which the cells themselves and their connections are liable. When you have determined up to which cell the battery works, carefully examine from that point if any connection is loose, or corroded; if any cell leaks, or works irregularly, (salts creeping up, zinc being eaten away, etc.). All this of course requires some practical knowledge of electrical matters in general, as well as of the details of the battery used, but the labour expended in acquiring this knowledge is sure to repay amply in time. Many disappointments will be avoided and time and money saved by him who can dispense with external assistance in the management of these little matters.*

AMALGAMATION.

Unless a piece of zinc be perfectly pure and of perfectly homogeneous consistence, it acts, when plunged into dilute acid, as two or more metals, and "secondary actions" are set up at its surface. That is to say the points where the zinc is harder, or contains iron or arsenic (as all commercial zinc does) act as negative electrodes to points where the metal is softer, and currents are set up between such points through the liquid and the plate, and thus the metal is gradually eaten away. "Amalgamation" obviates this serious drawback by giving to the surface of the metal a homogeneous consistency. The modus operandi is as follows: wash the zinc with dilute sulphuric acid (1:4), then drop a little mercury upon it, over a plate, and rub in the adhering drops until the whole surface is uniformly shining and smooth. A hissing noise heard when the plates are immersed in the acid and due to the evolution of hydrogen is a sign that the zincs require re-amalgamation.

* Some readers of my first edition suggested that I should have mentioned the price of the various apparatus mentioned in their book. This information they will obtain by applying to Mr. Schoth, Electrician, 232 Euston Road, London, who will give every information, and is prepared to supply the batteries and accessories described in this volume.

THE PEROXIDE OF MANGANESE ELEMENTS.

A Leclanché cell (fig. 34) consists of an outer vessel made of

FIG. 34.

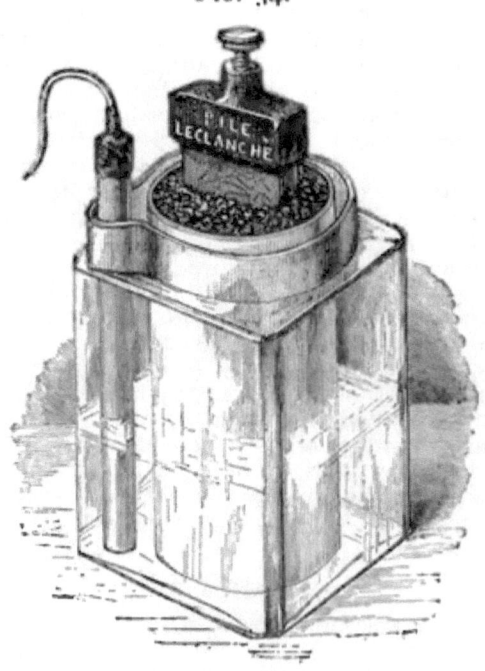

Leclanché cell; old pattern. To the left a rod of zinc; in the middle a porous vessel containing pyrolusite in fragments and a plate of carbon. The external jar is made of glass, and is seen to be half filled with solution of ammonic chloride.

glass or of vulcanite, containing 1st, a rod of zinc and a plate of carbon as electrodes; 2nd, peroxide of manganese, as depolariser; 3rd, saturated solution of chloride of ammonium, as exciting liquid. The advantages of such a combination is that the zinc is not consumed when the cell is not in circuit. The depolarising power of the pyrolusite is quite sufficient when currents of short duration (as for feeding an induction coil), or passing through high resistances (such as that of the human body) only are required.

Its electromotive force, when freshly charged, being 1·5 volt nearly, is higher than that of the chloride of silver and sulphate of copper cells. Owing to this fact 30 Leclanchés are equivalent to 40 of the weaker elements: the number of cells may thus be reduced, which is an advantage when bulk and weight enter into consideration, as in portable batteries.

The internal resistance of a Leclanché is somewhat less than that of a Daniell of the same size.

The chemical changes which accompany the evolution of electricity, are that $2n$, $2NH_4Cl$ and $2MnO_3$, yield $ZnCl_2$, H_2O, $2NH_3$, and Mn_2O_3; the peroxide of manganese is reduced, water is formed, together with ammonia and chloride of zinc.

Though theoretically no chemical changes whatever should take place in a Leclanché cell which is not in actual use, yet in practice it is found that such is not the case, especially in the small sized elements made for medical batteries. This is due to the fact that the ingredients used are not perfectly pure. The zincs should be drawn instead of cast so as to secure the utmost homogeneity, and be thoroughly amalgamated. The sal ammoniac should be chemically pure, and the pyrolusite chosen of the best sort, in needles, not powdered. There occurs sometimes, notwithstanding all precautions, a deposition of crystals of oxychloride of zinc and double chloride of zinc and ammonium on the electrodes, which interferes eventually with the activity of the cell, by preventing the necessary chemical action. This occurs chiefly in the medical elements, which contain but a small quantity of liquid. These deposits are not very soluble in water, but are more so in a saturated solution of chloride of ammonium. Hence it is necessary to keep a slight excess of this salt in the cell; too considerable an excess is to be avoided as giving itself rise to crystallization. The crystals once formed are to be removed by scraping the electrodes.

The latest model of the Leclanché element differs from the pristine type in the disposition given to its constituents. Formerly the carbon was contained in a porous vessel, and surrounded with fragments of pyrolusite more or less tightly packed in that vessel. Now the pyrolusite is moulded under very high temperature ($100°$ C.) and great hydraulic pressure (300 atmospheres), and with the admixture of powdered carbon and shellac, into thick plates. This improvement has not only enhanced the value of the Leclanché as a constant element, but has considerably simplified the manipulation required for the changing and cleaning of the cells. The zinc, carbon, and depolarising plates are simply bound together by means of an elastic band, and deposited in the vessel half filled with the solution.

The following instances will give an idea of the extraordinary durability of Leclanché elements (of large size) when in the hands of intelligent persons. A battery at a railway station has been in use from July 1867 till August 1876, at which time the zincs were still fit for work. During those nine years it was never repaired; chloride of ammonium was added only once, and water occasionally added to make up for the loss by evaporation. Another battery worked a signal bell continuously for 23 hours a day during 11 months without being touched.

In the Leclanché, as in many other cells, a source of danger is the creeping of salts upwards to the connections. Thorough protection of the latter by paraffin or varnish, and the smearing of the upper part of the vessel with grease are essential to their preservation.

I may add that though pure or amalgamated zinc should theoretically have no advantage over crude commercial zinc in working a Leclanché, yet owing to the tendency to crystallization upon the zinc rod, it is advisable to have the latter thoroughly amalgamated, and even if possible to keep its lower extremity surrounded by a few drops of mercury at the bottom of the cell. Much care should also be bestowed upon the selection of the ammonia salt—many samples of which contain impurities which lead to a speedy arrest of the cell-activity.

A modified Leclanché has been constructed and described by Prof. von Beetz about twelve years ago, in which the carbon was replaced with a piece of platinum. The electromotive force of this combination was found to be somewhat higher than that of its prototype. Prof. von Ziemssen spoke very highly of it, but ulterior experience has not justified the hopes raised by the new element. It consisted of a test tube through the bottom of which the platinode passed. The zincode, fixed to a cork plunged in the solution of sal ammoniac with which the tube was filled. The peroxide of manganese occupied about the lower third of the tube.

Fig. 35.

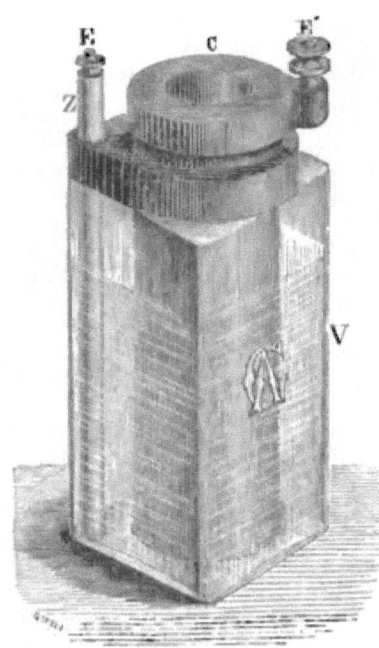

Gaiffe's peroxide of manganese cell. V, glass vessel containing a solution of chloride of zinc, in which are immersed a rod of zinc Z; and a cylinder of carbon hollowed out to receive the pyrolusite. E E' screws for fixing the connecting wires.

Coxeter, who has shown the most praiseworthy zeal in his endeavours to improve medico-electrical apparatus, has constructed a zinc-platinum element on the same general pattern as the familiar medical Leclanché. His batteries are light and compact; but I have had no personal experience as to their durability. The cells are not separable, but moulded in a mass of vulcanite poured into the case of the battery, and covered with a layer of resinous material. This arrangement may diminish the cost of production but makes the recharging a difficult matter.

Gaiffe, of Paris, has made two alterations in the Leclanché cell, (fig. 35). He has substituted a solution of chloride of zinc for that of the ammonia salt, and has placed the manganese inside the carbon which is hollowed out for the purpose. The electro-motive force of this cell is 1·3 volt. The chloride of zinc has the advantage of being more soluble, and does not so readily form the "creeping salt deposits" which frequently endanger the internal connections of a battery. The pyrolusite is very easily replaced when exhausted—at least in the case of the large-sized cells. In the smaller ones, on the contrary, the danger is that it tends to solidify into a hard mass if left too long unchanged; recharging becomes then impossible.

THE CHLORIDE OF SILVER ELEMENT.

Zinc and silver are the metals used as electrodes; chloride of silver is the depolarising agent. The electromotive is force equal to 1 volt nearly.

A chloride of silver element as usually made consists of a wide, flat-bottomed, test tube filled with dilute acid. The stopper consists of paraffin into which the electrodes (a rod of zinc and wire of silver) are fixed. The chloride of silver is cast, in the shape of a solid cylinder, around the silver wire. These cells work exceedingly well, as testified by Mr. Warren De La Rue who has fitted up a battery of 12,000 for scientific purposes.

Gaiffe of Paris, has devoted much attention to the improvement of the chloride of silver element, and brought it to a very high state of perfection with reference to its portability. A cylindrical box of vulcanite, with a top that screws down hermetically, contains a plate of zinc, and a wire or plate of silver surrounded with fused chloride of silver. The zinc and silver are separated by a pad of bibulous paper, which, when moistened with a weak solution of chloride of zinc contains all the fluid necessary to work the cell to exhaustion. A reference to the figures will make the details of the construction clear.

The cells are made of two sizes; the large V is only $3\frac{1}{2}$ inches long by $1\frac{3}{8}$ in diameter, and weighs but two ounces; it is used for working coils; etc. The smaller size is mainly used for galvanic batteries. The experience I have had of these elements is most

GAIFFE'S CHLORIDE OF SILVER ELEMENT.

FIG. 36. FIG. 37.

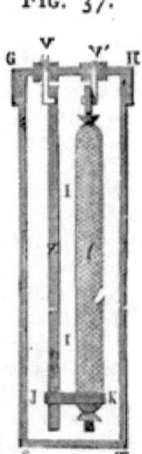

G, H, S, T, vulcanite cell. The top screws on, and carries two metallic pieces V, V', forming the poles to which are fixed, Z, a plate of zinc, and Y, which in fig 36 consists of a shallow silver cup filled with chloride of silver. The two are separated by a pad of moistened bibulous paper, I; the whole being fastened with an elastic band, J, K.

In the other form (fig. 37), the silver electrode is in the shape of a rod, surrounded with fused chloride of silver. The whole is contained in a muslin bag, Y.

In order to recharge the cell, unscrew the top of the cell, remove the elastic band, disconnect the zinc and silver electrodes, renew the chloride of silver (of which it is necessary to have a supply ready moulded), put in a fresh zinc and pad of blotting paper, moistened in three per cent solution of chloride of zinc. Secure with the India-rubber band, dip the whole in the exciting liquid, and fasten the lid. The whole operation takes but a very few minutes. If the cell remains long unused, the zinc is liable to become incrusted with oxide; it is then necessary to scrape its surface and moisten the pad with a few drops of water.

favourable. Their electromotive force is about the same as the Daniell's, 1·03 volt and they give a very constant current. As regards convenience they are far by the most portable cells ever constructed, and last a long time without requiring cleaning or recharging, which operation is very simple, and can be carried out by any intelligent person. The only drawbacks of the Gaiffe cells are their elevated initial cost (the working is not very dear, all the reduced silver being recovered) and unless used frequently, the incrustation of the zinc with inert oxide—to which may be added the necessity of obtaining the recharging materials from the maker abroad.

M. Gaiffe has made me a battery of 60 such elements, of which I can speak in terms of the highest praise. It has stood during many months the double test of periods of great activity, and of travels of more than 2,000 miles, during part of which it shared

FIG. 38.

GAIFFE'S CHLORIDE OF SILVER BATTERY.

G, Galvanometer fixed on the lid, with its conducting wires. The connection is established through the hinges. The advantage of this arrangement is, that when the battery is shut, the needle does not rest upon the pin, and does not wear it out uselessly.

The element board carries a double dial collector with the handles, M, M'. The metallic studs are marked, 0, 2, 4, etc. The rheophores are attached at B', B'. V, V, V, V, screws fastening down the element board. When not in use the handles rest on the dark studs marked "Repos." The commutator is left out; the interruptor marked I.

The cells F, F, etc. are arranged in rows, into trays. The positive and negative poles of each cell, H, H, etc. protrude through the upper part of the tray which is placed standing in the box. The inferior surface of the element board carries a number of springs, each of which presses upon one pole, and so the connection is established between the cells and the dial studs. A drawer at the bottom of the battery contains the accessories.

Size of 60-celled battery : 13 × 7 × 9 inches; weight, 15 pounds.

the fate of ordinary luggage, without giving any signs of becoming weaker, or of being otherwise out of order. I am not aware of any other battery which combines to the same extent, durability, portability, and efficiency. At the same time the cell is not likely to be much used except by those who are in the daily habit of applying electricity, owing to the rather serious drawbacks just mentioned.

One feature of the battery which I do not approve of is the use of springs to establish connection between the cells and the dial; they unnecessarily add to the weight of the instrument and are very difficult to manage oneself when some alteration becomes necessary within.

[I am at present experimenting with a modified chloride of silver battery, made at my instance by Schoth, of 232 Euston Road, London. Should the new elements, continue to fulfil my expectations, as they have done hitherto, electrotherapeutists will be placed in possession of by far the most portable battery ever made. The cost of the cells is not greater than that of the ordinary Leclanché. They appear to be free of the incrustations, which interfere so much in the working of Gaiffe's model. Finally the recharging can be carried out by the owner, being simple, cleanly, and rapid.]

THE SULPHATE OF COPPER ELEMENT.

THE original form of this element is the well-known Daniell, the most extensively used of all cells. It differs from the Bunsen (fig. 43), by the substitution of copper, and a solution of sulphate of copper for carbon and nitric acid respectively. The hydrogen on meeting the solution precipitates metallic copper and sulphuric acid is formed.

Its electromotive force is 1·08 volt when the sulphuric acid is diluted 4 times by weight.

Many modifications in the details of its construction have been devised to meet special ends. The following have been used in electro-therapeutics.

The Siemens-Halske cell has in this country a historical rather than a practical interest. The porous diaphragm is replaced by a thick layer of paper-pulp; saw-dust, sand, felt, &c., have also been used. The resistance is thereby greatly increased and the durability of the cell notably improved, but the sulphate sooner or later finds its way to the zinc which necessitates periodic cleanings and refillings. Remak, in Germany, to whose efforts and influence we owe the restoration of the galvanic current into medicine, made the greater part of his experiments with this element.

In order to provide a continuous supply of sulphate of copper, several contrivances have been imagined; in the Meidinger cell for instance, a vessel filled with crystals and with its mouth downwards is disposed as in the figure.

The Callaud or gravitation cell does not contain any diaphragm. The superior specific gravity of the saturated solution of sulphate of copper as compared with that of sulphate of zinc is taken advantage of, and the two liquids are simply superposed. Of course this precludes the possibility of using the cell for any but stable batteries. The chief merit of these elements is their cheapness and simplicity of structure.

SULPHATE OF COPPER ELEMENT. 63

FIG. 39.

Meidinger element. B, is a glass vessel, with a neck dipping into the porous vessel containing the copper electrode. The vessel is surrounded by a large circular zinc. The crystals and saturated solution with which it is filled before being inverted into the cell, keep up the supply of sulphate of copper constantly renewed.

FIG. 40.

Callaud element. A copper wire C, encased in a glass tube in its middle, and bent into a proper shape at its lower portion, stands in the middle of the cell. A plate of zinc is cut and bent so as to fit in the upper part, water is then poured to the requisite level, and crystals of sulphate of copper dropped in. In virtue of its gravity the inferior layers of the liquid—a saturated solution of the sulphate—remains distinct from the upper one into which dips the zinc.

THE SULPHATE OF MERCURY ELEMENT.

This element, also called after the name of its inventor Marié-Davy, consists usually of a zinc-carbon pair, the carbon extending to the bottom of the cell and dipping into a layer of bisulphate of mercury. The cell is filled with water which slowly dissolves the salt. Metallic mercury is thrown down when action is set up.

FIG. 41.   FIG. 42.

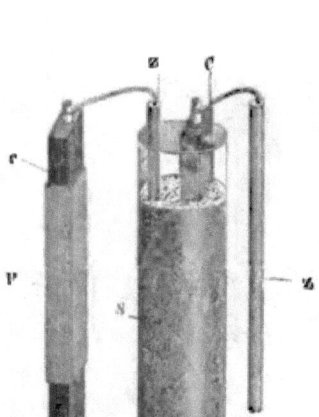

Gaiffe's sulphate of mercury cell. A zinc rod and a plate of carbon C, dip in a large test tube filled with saw-dust, S, moistened with acidulated water. At the bottom there is a thick layer of sulphate of mercury.

Trouvé's hermetical element. The upper half of the vulcanite cell is occupied by the circular carbon and rod of zinc as shown in the figure. The lower half is filled with a solution of bisulphate of mercury. When now the cell is placed on the side, the liquid comes into contact with the zinc and carbon, and enters into action.

The cell is very constant when worked through high resistances, and its electromotive force is the same as the Leclanchés; but it is exposed to the creeping of salts upwards, and is not so clean as the Leclanché.

Trouvé, of Paris, has made an ingenious portable mercury cell which enters into action only when turned upside down. The cell is hermetical, and very well adapted for working induction coils.

Gaiffe's cell consists of a shallow trough of vulcanite with a piece of carbon at the bottom. Some sulphate of mercury is spread over it, moistened with water, and the zinc in the form of a lid laid over the whole. It is extremely convenient for the purpose of working coils, and is figured in the paragraph on "Faradism."

THE BUNSEN AND GROVE ELEMENTS.

THE Bunsen element, consists of an earthenware vessel filled with dilute sulphuric acid into which is plunged a cylinder of amalga-

Bunsen element. Consisting of an external earthenware vessel which contains the zinc, Z, in the shape of a cylinder, bathed in dilute sulphuric acid. Within the zinc stands a porous vessel holding the carbon immersed in strong nitric acid.

mated zinc. Within the cylinder stands a porous cell of baked clay filled with strong nitric acid, and containing a prism of carbon. The action of a Bunsen is as follows:—the zinc is attacked by the sulphuric acid, sulphate of zinc being formed. The oxygen formed by the electrolysis of the water remains about the zinc, whilst the hydrogen combines with some of the oxygen of the acid which is reduced to nitrous acid, as evidenced by the evolution of the characteristic red fumes. As long then as there is a sufficient quantity of acid to oxidise the hydrogen, a constant current circulates.

Grove's cell is essentially the same as Bunsen's, the main difference being the replacement of carbon by platinum. Its chemical reactions are the same. The currents supplied by these two cells are very powerful, their electromotive force being about 2 volts, and their internal resistance being capable of being reduced to a fraction of an ohm. Once charged, they work very regularly for a few hours. But the unpleasant fumes they evolve, and the difficulty of manipulating the strong acids restricts their use considerably. Their application in electrotherapeutics is limited to the heating of wires for the galvanocautery.

SINGLE FLUID ELEMENTS.

Zinc-platinum element.—This element, known as Smee's, consists of a plate of zinc, and a plate of platinised silver dipping in dilute sulphuric acid. Platinised silver is used on account of the less tendency hydrogen has to accumulate upon a surface of finely divided metal. In Frommhold's battery platinised lead is used, being cheaper. The electromotive force of the Smee is under 1 volt.

The zinc-carbon element.—This element which has been popularised in the medical world by Stöhrer, consists of a pair of zinc and carbon plates dipping into a weak sulphuric acid solution (1 in 10 or 20 parts of water). Its electromotive force is higher than the Smee's (1·4 volt); and like platinised silver, the carbon, owing to its rough surface, offers a certain opposition to the adhesion of hydrogen bubbles, and thus polarisation is somewhat delayed.

Fig. 44.

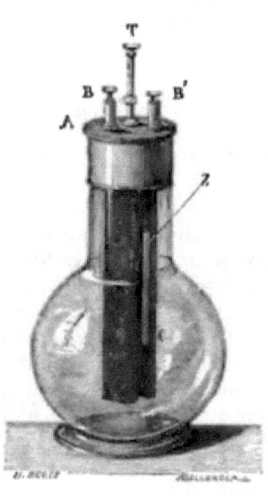

The Grenet element. A plate of zinc, Z, between two carbons, C, C, dips into a solution of bichromate of potash in sulphuric acid, 1 part, and water 10-20 parts. A, vulcanite plate bearing the binding screws; B, B', for connection. By means of the rod T, the zinc is withdrawn from the liquid when the cell is not in use.

SINGLE FLUID ELEMENTS.

The bichromate of potash or Grenet element.—This is a zinc-carbon element in which the addition of bichromate of potash to the exciting liquid gives a higher electromotive force, and a higher constancy, as it acts as a partial depolariser. The Grenet is largely used for working induction coils and is well adapted to this purpose, as it gives a very powerful current for a moderate length of time, and has a very low internal resistance. In Stöhrer's induction apparatus, chromic acid replaces the bichromate of the usual Grenet. Electromotive force equal to about 2 volts.

The zinc-carbon battery with the addition of bichromate of potash or of sulphate of mercury is a convenient one for the purposes of electrolysis of tumours, (fig. 45). Large single fluid elements have also successfully been used for galvano-caustic purposes.

FIG. 45.

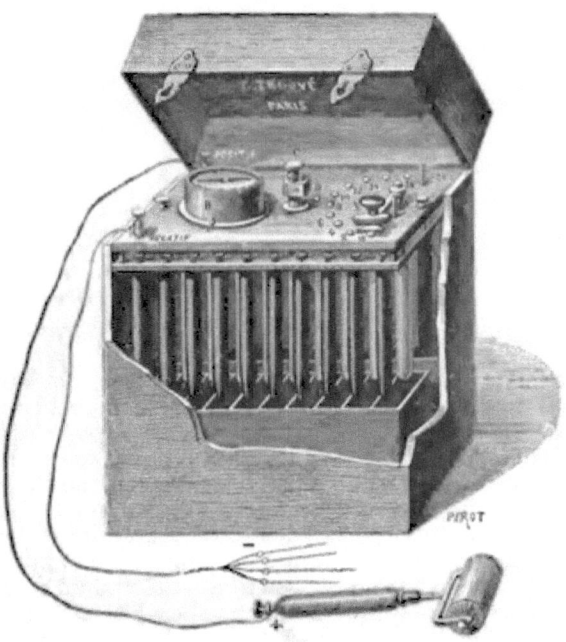

Single fluid battery. It consists of 20 zinc-carbon elements taken (by twos) by means of a dial collector A; B, galvanometer; C, D, arrangement for the gradual immersion of the plates. To the terminals are attached the rheophores carrying needles and a roller electrode such as is used in electrolysis.

POLARISATION CELLS.

WHEN two plates of platinum are dipped into water and attached to the poles of a battery they become polarised: that is the anodic

FIG. 46.

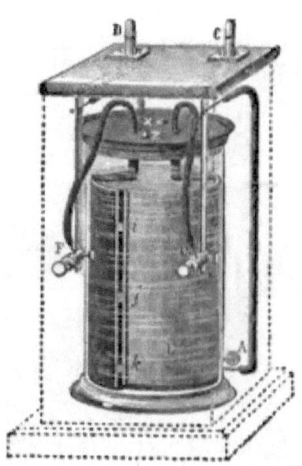

plate becomes covered with bubbles of oxygen, the kathodic with bubbles of hydrogen. If now the plates are connected with the terminals of a galvanometer, it is found that a polarisation current is set up from the hydrogen through the liquid to the oxygen (*i. e.* in the opposite direction to the battery current), and that this polarisation current may last a considerable time if the plates are large enough. This property has been made use of in the construction of what are now called secondary cells or accumulators. For instance in fig. 46 we have a representation of Planté's cell. It consists mainly of two plates of lead, L, rolled into spirals, one within the other, but kept from contact by means of strips of vulcanite, $i\,j\,k$, they are connected to the polarising battery by means of two wires E F, and to two other terminals, C D, which give the polarisation current. To set the battery in action, we have simply to fill the secondary cell with water containing 10 per cent. of sulphuric acid, and connect the binding screws, E F, to a battery of 4 Daniells or 2 Bunsens. After a while the cell is charged, when if a wire be attached to C and D, the polarisation current flows through it and keeps it hot until the electromotive force stored up is expended.

B. FARADISM.

About fifty years ago, Faraday discovered that if a magnet was moved in the neighbourhood of a closed circuit, such as a ring made of metal, a current of electricity was generated in that circuit; he also found that a galvanic current which varied in its strength had a similar effect upon a neighbouring circuit. Hence the name of "faradic" applied to such currents, also known as "induced" from their peculiar mode of generation.

Induction is the the name given to certain phenomena exhibited by a wire or other conductor in the neighbourhood of which a current or a magnet is placed: as long as no change occurs in the distance or strength of the latter, nothing occurs in the wire; but as soon as any such change is set up, and as long as it lasts, electromotive force is generated in the wire. The sudden magnetisation of a piece of iron and the sudden making of a galvanic current in a wire, may be considered as equivalent to the sudden bringing of the magnet or circuit from an infinite distance, to the spot they occupy. Conversely sudden demagnetisation and sudden breaking of current, are equivalent to their sudden removal from that spot to an infinite distance. Likewise, sudden changes in the intensity of a current or magnet, correspond to sudden changes in the distance of that current or magnet from the circuit they influence; every increase acts as an approximation, every diminution as a removal. Gradual changes in the intensity act as gradual changes in the distance of the magnet or current. Hence we may say that whenever a current (or magnet) is (suddenly or gradually) made or broken near a wire (that is, whenever it is brought from an infinite distance into its neighbourhood, or thence removed to an infinite distance), or more generally that whenever any variation of strength or position occurs in the current (or magnet), a certain amount of electromotive force is induced in the wire.

This amount is absolutely definite in each case, depending upon a number of circumstances such as the length of the wire influenced and the strength of the current, the distance between the wires, etc. Suppose that in one case a current of 1 unit be made suddenly, that is in a very short time (say $\frac{1}{100}$th of a second), in the wire AB; and in another case be gradually (say in one second) made, that is to say, increased from 0 to 1, we shall have in both instances a total electromotive force X generated in CD. But it is evident that at any instant of its existence the second current is 100 times weaker than the first, and though the same total quantity of electricity is conveyed in the two cases, the physiological effects in the first will be much greater owing to the concentration of that quantity in a very short time.

EXPERIMENT XX.—Connect the wire CD with a galvanometer, and produce changes in AB. You will find that the needle is not always deflected in the same direction. When the current from B to A is made, increased, or brought near, the current induced in CD is ob-

Fig. 47.

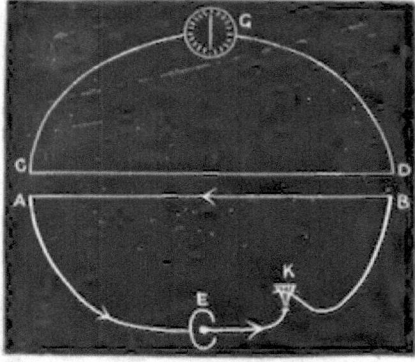

served to flow in the opposite direction, viz. from C to D. The reverse takes place when the current BA is broken, diminished or removed. If then by means of a key, K, we alternately make and break the current, the needle alternately swings to the right and to the left as the case may be. Similar phenomena are observed if AB is a magnet subjected to corresponding changes.

The amount of electricity induced in CD is proportional to the strength of the current in AB, and to the proximity of the two wires.

Fig. 48.

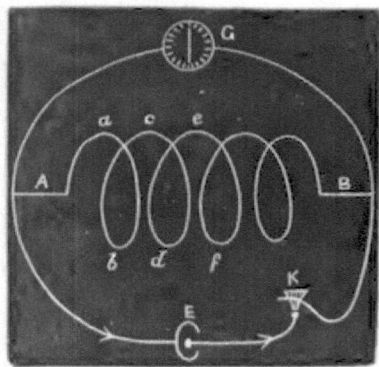

EXPERIMENT XXI.—Let the wire AB instead of straight be made into a solenoid or coil. It is evident that the consecutive turns may be considered as constituting a series of parallel wires (a, b, c, parallel to c, d, e, etc.). You will find that every variation of the current in its passage through each turn from B to A does induce electromotive force in the contiguous turns. An induced current is thus generated in AB which obeys the same laws as that arising in an independent wire which is opposed in direction to the primitive battery current, when the latter is made or increased, similar in direction when the latter is broken or diminished. This current is designated as the primary or extra-current.

Let us suppose (see diagram 48), that AB is connected with the two poles of a battery E, and that the current can be made and broken at will by means of a key, K. Connect AB with a galvanometer G. We now find that on breaking the current from the battery, a strong induced current is generated which causes the needle to be strongly deflected. But on making the current again no such deviation is observed; the induced current being opposed in direction to the battery current acts as a momentary resistance to it. The battery current is weakened and retarded in the process.*

EXPERIMENT XXII.—Make the second wire CD also assume the

FIG. 49.

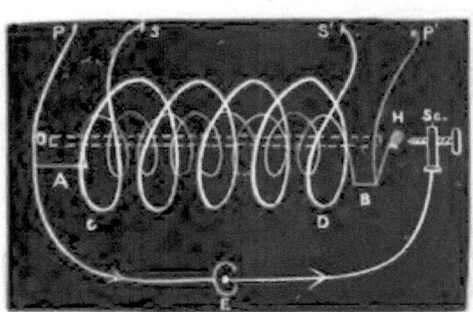

PP' in this diagram corresponds to A" A'" in fig. 50; ss' to aa'—AB to H—CD to H'—O to wires in O—Sc to P—H to stud under P— and junction of battery wires with A and Sc, to AA.'

shape of a coil, so as to surround the coil AB. It is evident that each coil turn in CD may be considered as parallel to each turn in AB. The total inductive influence of the coil AB upon the coil CD will be proportional to the total number of turns in the first multiplied by the total number in the second. By increasing the number of turns, you therefore increase the amount of electromotive force generated. Experiments will show you that this amount is independent of either the material or the thickness of the wire. These circumstances influence only the resistance in the coil.

The actual apparatus for demonstrating the facts of induction is shown in fig. 50. The wires of a galvanic battery are attached to AA', and a sensitive galvanometer to A" A'" or aa'. The current may be made or broken at P, and a magnet (or soft iron wires) inserted into or removed from O. It would lead us too far to describe in detail the delicate experiments required to prove all the

---

* The experimental proof of all these facts is not so simple as would appear from the diagram and description here given. (Cf. Gavarret, *Electricité*, vol. ii., p. 186 ff).

FIG. 50.

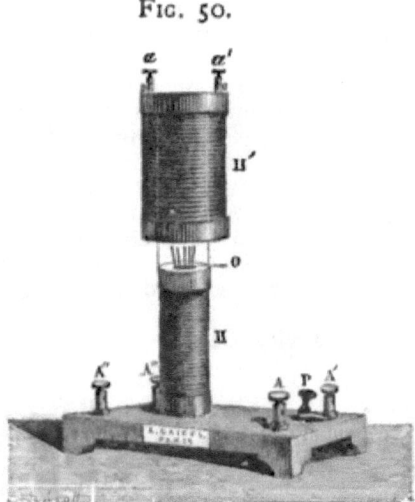

Apparatus for demonstrating the phenomena of induction: H primary coil, the two extremities of which are connected (1) with A″ and A‴; and (2) with the stud under P and A. P is a spring which when pressed down connects the stud with A′. O is an empty space in the coil for the introduction of soft iron rods, or a magnet. H′, secondary coil, the extremities of which are connected with the binding screws aa'. The dotted lines show that H′ is hollow and may be made to include H.

The wires from the battery are attached to AA′ and the current through H alternately made and broken by means of the key P. The extra-current is obtained at A″ A‴, and the secondary at aa'.

The wires composing the coils H, H′ are spun over with fine silk dipped in paraffin, so as to insulate each turn from its neighbours.

statements concerning induced currents, the demonstration of which, owing to the very small quantity of electricity they convey, requires a very sensitive galvanometer (fig. 16).

The effects of a more or less rapid motion of a wire in presence of a current or magnet are observed by closing the circuit through H or placing a magnet in O, and bringing H′ so as to enclose H, and removing it again. The galvanometer shows that at each event a current is generated in H. Currents are likewise generated in it when H′ being placed so as to enclose H, the circuit through H is made, and when the soft iron core in O is magnetized—and when the circuit is broken or the soft iron demagnetized.

The direction of the current in H′ is the same when H′ is brought near, when the circuit through H is made, and when the core is magnetized—and opposed in direction to that generated when H′ is removed, when the circuit is broken, and when the core is demagnetised.

The currents in H′ grow with the strength of the rapidity of the motion, the strength of the current in H, and of the intensity of the magnetism in O. They grow also with number of turns in H and H′ (allowing for the increased resistance) and with the amount of overlapping of H by H′.

The ordinary medical induction apparatus consists of the parts shown in diagram, fig. 49.

1. The battery E, composed of one or two elements, one of the poles of which is connected with the screw of the interruptor, Sc., the other with one end of the primary coil, A.

2. The automatic interruptor consists of the spring hammer, H, which when at rest is in contact with the screw, Sc just mentioned. Its free extremity impinges upon the iron core of the primary coil, and its fixed base is connected with the other end of the primary coil B.

3. The primary coil, AB, is made of a short thick wire. Its extremities are fixed to the battery and hammer respectively, as just mentioned. (In the diagram AB is drawn as a thin, CD as a thick wire).

4. A core, O, composed of a bundle of iron rods, inside the primary coil.

5. The secondary coil, CD, made of a long thin wire. It is entirely independent from the rest of the apparatus, and its two extremities are connected with two binding screws, s, s, to which the rheophores are attached.

6. Two wires branching from the extremities of the primary coil, AB, and terminating at two other binding screws, P, P', for the rheophores. These wires convey the extra-current.

7. All apparatus are fitted with contrivances for graduating the current strength.

The action of such an apparatus, is as follows:—The moment the current begins to flow in the primary coil, the iron becomes magnetized, the hammer is attracted, and the contact between the screw and hammer broken. The current ceases to flow. Immediately the iron is demagnetized, the attraction ceases and the hammer flies back, in virtue of the elasticity of the spring, into contact with the screw. Once more the current begins to flow and the same succession of phenomena is repeated. It is evident that the rapidity of these makes and breaks depends upon the time taken by the hammer to swing to and fro, that is upon the freedom of play of the spring. Provision is made in most apparatus to modify the rapidity of vibration by screw arrangements, etc., but usually this is quite insufficient for obtaining slow interruptions, which have then to be made by moving the hammer with the finger, or more conveniently by means of a simple spring key, (P, fig. 50) or of some special contrivance (fig. 53).

At each make of the battery current, then, we have two induced currents, one in each coil, opposed to it in direction; at each break we also have two induced currents, flowing in the same direction as the battery current.

We have seen that the strength of the induced current, bears a certain proportion to the distance between the two wires. When the secondary coil is made moveable, the current rapidly diminishes when that coil is removed, so as to include less and less of the primary. In many apparatus (Du Bois-Reymond's, Stöhrer's, etc.), the secondary current is graduated by an adaptation of this principle. (See also fig. 50).

In other instruments the two coils are immovable, the secondary entirely surrounding the primary. Both currents are then usually graduated by means of a copper cylinder which is made to include, at will, a portion of the iron core. The reader is referred to the larger treatises, for an explanation of the mode of action of this contrivance. Suffice here to say, that the cylinder delays the inductive process, and so weakens both primary and secondary currents. The greater the length of the core it includes, the greater its effect in that direction.

In some instruments the wires are tapped at different lengths which may be taken successively by means of a dial collector: the current is proportional to the length so used. Or again, a water rheostat is included in the circuit: this is a very effectual mode of graduating the strength of faradic currents, but suitable only for the cases where graduation need only be empirical. The same may be said of the damping process, by means of the cylinder. Both are suitable for treatment; but for diagnosis the arrangement with a sliding secondary coil alone gives us the means of establishing comparisons between the current strengths employed.

There exist, unfortunately, no means of measuring the strength of faradic currents in medical practice. The galvanometers necessary for this purpose are very expensive and only indicate the quantity of electricity conveyed, whilst the physiological action of the currents depends upon another factor, viz., the suddenness of their rise and fall which escapes almost every possibility of estimation. In order therefore to obtain currents comparable among themselves, instruments of exactly the same construction should be used, supplied by galvanic elements of perfectly constant value. The readings along the scale would not even then be proportional to the strength of the currents generated in the secondary coil, for as the latter is made to include a greater length of the primary, the currents increase in a geometrical and not in an arithmetical progression.

Induction apparatus (cf. fig. 51) are usually graduated in centimetres, the zero being placed so as to indicate that there is "no distance" between the primary and secondary coil, $i.e.$, that the primary is entirely included within the secondary. As the former is moved along the scale (see graduation in fig. 51, indistinctly marked with numbers; 0, 5, 10 ... 30), its position is indicated by saying that the coils stood at 5, 10, 20, or 30 cm. distance from one another.

EXPERIMENT XXIII.—Take an induction apparatus (fig. 51) and connect the terminals of the secondary coil with a multiplying galvanometer, and note the deflections when the secondary coil stands at various distances from the primary. You will find that the strength of the current increases very rapidly as the distance diminishes. The actual numbers will differ from instrument to instrument (unless the latter are absolutely alike). The following were found in actual experiments:

Distance of coils in cm.		50	45	40	35	30	25	20	15	10	5	0
Strength of current in degrees of needle deflection.	Exp. I.	3·5	5	7·5	11·5	19·5	33	51	71	95	110	122
	Exp. II.	11	12·5	15	18	22·5	28·5	36	52	65	92	135

We have seen that the make primary current simply retards the development of the primitive (or battery) current, acting as a temporary resistance in the primary coil. The result is that the time taken by the battery current to attain its maximum strength is prolonged; hence, the generation of electromotive force in the secondary coil is protracted, and as a consequence, the make secondary current is weakened in proportion to its longer duration.

The break primary current (or extra-current), and the break secondary current are much more rapidly induced, the whole electromotive force generated coming into play instantaneously, so to speak; hence they alone have an appreciable physiological effect when coils of the usual power are used and the human body is included in the circuit. The extra (or primary break) current manifests itself as the spark which flies between the hammer and screw. It is due to the passage of the current through the air as the hammer recedes from the screw. Hence it is most vivid when the extra-circuit is not closed, and becomes fainter as the resistance in the extra-circuit (P to P', fig. 49) becomes smaller.

Hitherto we have said nothing of the influence of the core of soft iron in the primary coil upon the induced currents. It becomes magnetised when the battery current begins to flow through the primary coil, remains magnetised during the whole time of the flow, and ceases to be magnetised when the current is broken. As previously mentioned, the influence of a magnet made or unmade in presence of a wire (brought near or removed from a wire) is precisely the same as that of a galvanic current. The magnetisation and demagnetisation of the soft iron increases the amount of electricity generated in the coils by the make and break of the battery current; for as stated, this magnetisation and demagnetisation occur synchronously with make and break, and induce electromotive force in the same direction as make and break respectively.

The physiological distinction of poles has to be observed in faradic as well as in galvanic currents; (1). This is obviously the case with the primary coil where the make current is neutralised by the battery current and does not flow through the body. (2). The secondary make current, unless produced by a very powerful coil, has practically no influence whatever when circulating through the human body. The anode and kathode of the break current, therefore, determine the polarity of the terminals of the secondary as well as of the primary coil ($a\ a'$, $A''\ A'''$, fig. 50).

The difference between the primary and secondary current depends upon the fact that the former arises in a short thick wire, the latter in a long thin one. Ohm's law explains the phenomena observed. Suppose the electromotive force generated in the primary coil, the resistance of which is, say, 1 ohm, to be equal to 5 volts; that in the secondary (resistance 1000 ohms) to be 100 volts. If we send these two currents successively through a galvanometer of say 4 ohms resistance, the deflections of the needle indicate current strengths of:

In the first case: $\frac{5}{1+4} = 1.$

In the second case: $\frac{100}{1004} = \cdot 1$ nearly

If the resistance of the galvanometer be 50 ohms, the two currents will be almost equal, for $\frac{5}{50+1}$ and $\frac{100}{1000+50} = \cdot 1$ nearly.

If again the resistance in the circuit be 2000, as for instance that of a portion of the human body, the strength of the current in the first case will be only ·0025, in the second ·033, that is in the proportion of 1 to 13. These numbers are only typical of course, but serve to illustrate the principle just stated.

The difference between two such currents is commonly expressed by saying that they are of "low and high tension" respectively. What we have just stated concerning them will explain why the "high tension," or long fine wire, has to be used for exciting the dry skin, (which is a bad conductor); whereas the other is superior for the purpose of stimulating muscles through the wetted skin, where the resistance is much less. In the first case a small quantity of electricity only, concentrated at the end of a wire brush, is required; in the second a much larger quantity owing to the current-diffusion, but impelled by a weaker electromotive force owing to the less resistance to overcome, becomes necessary.

INDUCTION-APPARATUS.

FARADIC or induction-apparatus are of two kinds. In the first and now usual kind the generating power is an interrupted galvanic current; hence it is called "galvano-faradic." In the second, or magneto-faradic, a revolving magnet is the source of the current. The following remarks apply to the former only.

Two considerations are to be taken into account, in choosing an induction apparatus: Is it to be essentially portable? Is it to be used for diagnosis as well as for treatment?

Du Bois-Reymond's sledge faradic apparatus, with Gaiffe's improved contact breaker. The board A D is jointed at C for the sake of convenience. BB' are the connecting screws for the battery, (a couple of Leclanchés). L, lever which when raised as in the figure comes in contact with R, the vibrating hammer, and produces rapid interruptions. The slow ones are effected by pressing upon L so as to make it touch the pin R'.

For medical purposes the apparatus is supplied with 3 secondary coils, (the wires of which are of 66 metres, 1·4 mm.; 198m, 0·7 mm.; 600m, 0·225 mm. in length and diameter respectively).

When an instrument is for use at home only, a sledge apparatus (figs. 51 and 53) is the most satisfactory. It may be furnished with coils of various sizes of wires, with a slow interruptor; the cells which supply it are relegated to a convenient place, out of sight.

The utmost portability characterises some of Gaiffe's apparatus, such as figs. 54 and 55. Midway between these two kinds, stand apparatus fitted with Leclanché elements, (fig. 52).

Fig. 52.

Sledge faradic apparatus. Primary and secondary coils of the same length and size as those in the former apparatus.

EE'. Two Leclanché elements, connected by RR' with the coil and hammer.

L. Lever to set the apparatus into action, or to give single shocks at will.

h. Primary coil; HH' secondary coils of different sized wire. H' is in the compartment for the coil not in use.

O, O. Holes for the rheophores when the extra-current is wanted; for the intercalation of the pedal rheotome or of an external battery, if required.

There is no necessity for ordinary purposes, that the induction apparatus should be provided with elaborate contact breakers, for varying the rate of the interruptions. The ordinary interruptors vibrating 30 or more times a second, are sufficient. But for diagnosis, it is often necessary to give single shocks; this may be done by moving the hammer with the finger, but it is much better to have a spring-key attached to the instrument for that purpose (P, fig. 50). The advantage of Gaiffe's interruptor, (fig. 51), is that it may be used both for rapid and for single interruptions; it is at the same time the simplest and least liable to get out of order of all interruptors.

The next figure represents an induction apparatus similar to that shown in fig. 51 and provided with Gaiffe's new automatic interruptor. This contrivance allows to vary the frequency of interruptions from 1 to 60 per second by the simple movement of a lever, L. In the position shown it gives slow, at L' rapid interruptions; at L'' the instrument is at rest. The principle is that as the lever travels from L'' to L' the nicely balanced bar S I is made to rest more and more heavily upon the spring R, which is at the same time shortened by an ingenious mechanism. Now when the contact is made between these two pieces the current begins to flow through the electro-magnet E; I is attracted from R, the current is broken, I falls back and so on. It is obvious that the greater the length of R the greater the rebound of I; the shorter that length, the less that rebound. Hence the excursions of I become shorter and the vibrations quicker as L tends to approximate L'.

Fig. 53.

When an instrument has to be used for diagnosis it is essential that the physician should be able to graduate the strength of the current and record the strengths for comparison. To do this efficiently, we must have a moveable secondary coil, provided with a scale giving the strength of current in terms of the distance between the primary and secondary coil. This condition is fulfilled by the apparatus on the sledge principle, whether horizontal (figs. 51 to 54) or vertical as in the familiar Stöhrer's machine.

When an apparatus is designed for treatment only, the two coils may be immoveable and the current-strength modified by means of a draw tube (fig. 55).

The cells most generally used in England for working induction coils are Leclanchés, which are certainly preferable to the acid cells, and not much more bulky. If the owner of the apparatus has no one at hand to recharge his cells when run down, he will do well to have a couple always ready to slip into their place. An advantage of the sulphate of mercury cells (fig. 55), is that when a patient is receiving regular treatment a cell can easily be left at his house; and the coil only need then be brought there by the physician on his round. Chloride of silver cells, though perfect in their way, are not to be recommended for coils, unless intended for daily use, as after a while the deposition of oxide on the zinc interferes with their action, by increasing the internal resistance.

Though thick and thin wire coils do not differ in their fundamental physiological properties, yet their respective physical constitution endows them with very different effects. Thin long wire is necessary when "sparks" are indicated, as for the faradisation of the dry skin; a short but thick wire is best for stimulating nerve and muscle through the moistened epidermis. Hence the necessity of having two coils, when both purposes have to be fulfilled. In Gaiffe's sledge apparatus (figs. 52, 53, 54), two or three secondary coils made of different wire are provided. In the other kinds the primary coil (thick wire), as well as the secondary (thinner), is used according to the wants of the case.

For general purposes Gaiffe's apparatus (fig. 54), but of smaller size and fitted with sulphate of mercury, is probably the most convenient and effective. Apparatus fig. 53 fulfils all the higher requirements of the physician, whilst for treatment of the more ordinary kind and as a patient's battery, the pocket apparatus (fig. 55) or a simple primary coil with a Leclanché or other cell, will be found sufficient.

FIG. 54.

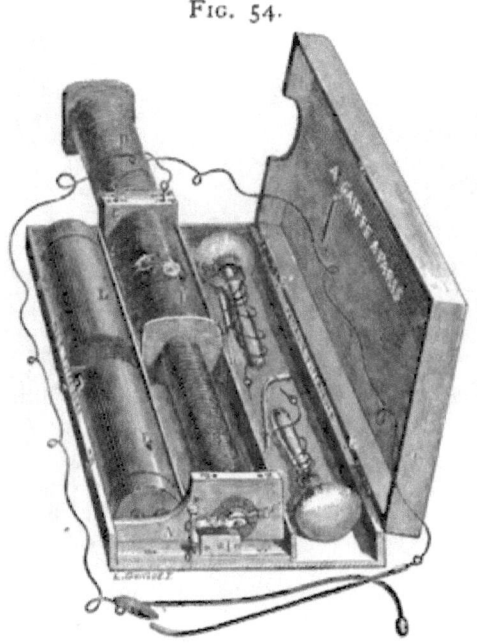

Gaiffe's portable faradic apparatus, with two induced coils.
Inducting coil; length, 25 metres; diameter 0 mm, 7.
Thick secondary; ,, 75 ,, ,, 0 mm, 7.
Thin secondary; ,, 450 ,, ,, 0 mm, 175.
LL. Two chloride of silver cells.
B. Primary coil.
H, H'. Secondary coils, reversable according to the one required.
C. Block for fixing the rheophores.
P, P'. Lever for regulating the interruptions, carried on the block I.
O. Button for making the slow interruptions with the finger.
E, E'. Holes for connecting the interrupting pedal; D, D', for an external battery, when required.
Size: length, 10 inches; width, 6 inches; height, 2½ inches; weight 3½ pounds.

A smaller model of this apparatus, with sulphate of mercury battery is made which combines the utmost portability and efficiency.

Defective action of a faradic apparatus may be due to several causes, having their seat in the various parts of the instrument, and is evidenced by the enfeebled vibration of the hammer.

The cells may be exhausted; and require cleaning and recharging. Some of the connections may be broken or oxidised. Some of the turns in the coil may be imperfectly insulated. But the most frequent part at fault is the interruptor. Sometimes a little manipulation of the screws which regulate its action is sufficient; or it is necessary to clean the platinum disk carried by the hammer (H fig. 49) at the point where it comes into contact with the screw (Sc.) and the tip of the screw itself. The reason for this is plain; oxide

FIG. 55.

Gaiffe's pocket faradic apparatus. M, coils. R, graduating tube. T, N, electrodes. P, hammer. E, block for the attachment of rheophores. The battery L consists of two cells formed of vulcanite troughs, a plate of carbon is let in at the bottom, and the zinc is in the shape of a moveable cover. In order to charge the battery, a spoonful of the mercury salt contained in the bottle K is put into each cell, with a little water. Once charged the battery works for about one hour. A smaller quantity only of the salt is needed when a shorter period of work is required.

collects at this spot, and offers a relatively considerable resistance to the passage of the battery current. This oxide is due to the chemical action of the battery current itself and of the break extra-current (spark) through the metallic surfaces, in presence of air. Careful scraping with a pen-knife, or scouring with fine emery paper, readily removes the oxide.

The advantage of a magneto-faradic apparatus is that it requires no cells to work it. On the other hand the labour of turning the handle, the noise it makes, and the roughness of its shocks, make, its use undesirable when a galvano-faradic apparatus can be had. It is a useful kind of instrument to keep in cases of emergency, when it is always sure to be found in working order.

A magneto-faradic machine consists mainly of an armature of soft iron revolving in front of the poles of a horseshoe magnet, and a pair of coils of insulated copper wire. These coils are fixed in Clarke's machine to the armature with which they revolve; in Duchenne's to the magnet itself.

In the first case electromotive force is induced in the coils by the successive magnetisation and demagnetisation of the armature due to the influence of the magnet in front of which it revolves.

Fig. 56.

Clarke's magneto-faradic apparatus, modified by Gaiffe so as to send the currents in the same direction, and allow of a more accurate regulation of the current strength.

A B B'. Horseshoe magnet.
H. Soft iron armature revolving in front of the magnet, and carrying 2 coils one of which only is visible.
M, R. Wheel and handle for making the armature revolve.
G. Regulator, articulated at O, which (according as it is pushed towards B or B'), increases or diminishes the strength of the current.
P P'. Supports of the apparatus.
M'. Receptacle for the handle.
C, C. Compartments for the electrodes.

In the second case induction is produced by the variations in the magnet itself due to influence of the armature.

In both these forms only the currents arising during the passage of the armature *from* the magnet are collected for use.

M. Gaiffe has improved the older forms of magneto-faradic machines inasmuch as he has combined the two systems and provided his apparatus with two pairs of coils, one fixed to the armature, the other to the coil. This increases the power of the instrument whilst it allows of diminishing its size considerably. He has in addition fixed a commutator to the axis of the armature, which combines the currents of the two pairs of coils and transmits them all in the same direction to the positive pole of the instrument.

The fact of a magneto-faradic apparatus giving a uni-direction current is a matter of considerable importance. For unlike the galvano-faradic make and break currents of which with the usual size of coils, the second has no appreciable physiological, and neither any chemical action, the to and fro magneto-currents are physiologically equivalent; and what is more they both have a sensible chemical effect.

ACCESSORIES.

Current Graduation—Rheostat and Collector.

The first condition to be fulfilled by a medical battery is to be furnished with ample means of graduating the current strength according to the requirements of every case. Looking at Ohm's formula $C = \frac{E}{R}$ we see that C can be modified, 1st, by making E, that is, the number of elements used, larger or smaller; 2nd, by making R, that is, the resistance in circuit, larger or smaller. There is yet a third method of graduating current-strength, that of establishing a derived current with a resistance variable at will.

A. The latter method has been adopted by many electro-therapeutists after Brenner's example. Its theory is explained under the head of Derived Currents, (p. 40). It yields a very fine graduation, but is very wasteful, and applicable only when the battery of Daniells is used. It as as far as I know, never used in this country. These reasons, and the fact that better results still are obtained by the use of a rheostat and absolute galvanometer in the main circuit, allow us to dismiss it without further notice.

B. In order to graduate the current by modifying the resistance in circuit, a liquid or wire rheostat is used, capable of interposing such resistance as to reduce the current to the minimum strengths usually required; by gradually diminishing the resistance, the strength of the current is gradually increased. For medical purposes wire rheostats, or resistance coils, are not suitable owing to their expense. Nor do they offer any marked advantage, since rapidity and fineness of graduation are requisite, rather than the interposition of *known* resistances. Liquid rheostats fulfil these conditions. They may consist simply of tubes filled with pure water, or dilute solution of salts, with a rod working like a piston as explained elsewhere. They need no graduation, the only datum requisite for introducing the necessary amount of resistance being the effect produced upon the strength of the current as indicated by the galvanometer. But they are not very convenient in practice, and the instruments made of graphite (cf. page 22) are likely to supply an important want in electrotherapeutics (fig. 57).

The advantages of such a rheostat as current regulator are:—First its simplicity and cheapness. It is obvious that the more or less

FIG. 57.

Medical Rheostat. BB, wires from battery; RR, wires to electrodes. Resistances are thrown in by turning the switch to the required position in the circle of studs.

complicated mechanism and multitude of wires necessitated by collectors being done away with, the battery will be less expensive and less likely to be out of order.

Second, the facility and fineness of graduation, especially if combined with a collector.

Third, the imparting to the current the qualities of "high tension." Upon this point I refer the reader to what has been already said under the head of "Constancy of the Current" (page 39).

Fourth, the possibility of estimating the resistances in circuit by the methods previously described (page 20), if the instrument is graduated.

One word of caution is necessary: when a rheostat is used, care must be taken before applying the current to the patient, that the full resistance be intercalated; otherwise an unpleasant or even dangerous shock might be given. The possibility of disregarding this rule has been held to be an objection to the use of the rheostat: but precisely the same might be said of any current regulator. In fact I have known the same accident to occur with a dial collector as happened to Duchenne with his water rheostat who sent the full current of a strong battery through a patient's head.

C. The third and usual method of regulating the current strength is by including in the circuit the number of cells requisite to bring the current up to the requisite strength. The simplest method of effecting this consists of connecting the electrodes (NP, fig. 60) directly to the poles of the cells by means of clips (A, B,) fixed to conducting wires. It is scarcely adapted to practice however; and all modern medical batteries are provided with a contrivance

Fig. 58.

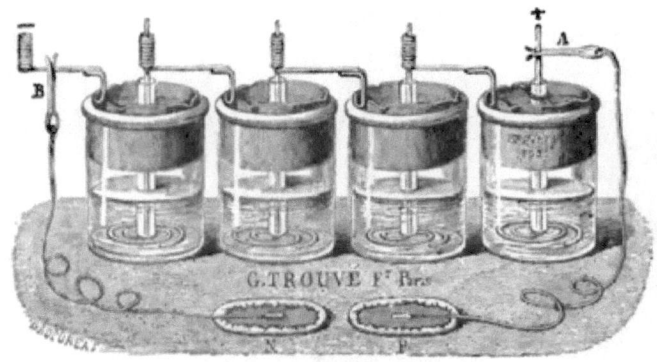

Four Callaud cells connected in series. NP, electrodes connected with wires to the clips AB, which may be fixed to any of the cells forming the battery.

known as a "collector" for the purpose of facilitating the manipulations required to alter the strength of the current as required.

In their general construction all collectors consist mainly of wires connecting the cells of which the battery is composed, to metallic pieces fixed on the element board. These metallic pieces are brought into relation with the binding screws for the rheophores by means of plugs, springs, etc.

Reference to the diagrams at page 37 and 87 will facilitate the understanding of the structure and mode of action of collectors.

Two essential conditions must be fulfilled by any collector intended for a practitioner's battery.

1. It must proceed by small increments in the number of cells. No collector should be made which does not allow taking cells by twos (if Leclanchés) or threes (if Daniells) *at most*. After the first 15 or 20, the increments may be larger.

2. It must not break the current with every change of current-strength. This is done by adopting precautions mentioned further on, when the various kinds of collectors are described.

The Plug Collector.

This is the simplest, the least likely to get out of order, but also the least convenient to work, of the various collectors. Several kinds of it exist, which will be found illustrated in fig. 32, where it will be seen that the wires from the cells are connected to metallic pieces, into which a pin or plug attached to the rheophore can be fixed.

Whenever such a collector is used it is often necessary to adopt some means allowing of changing the current strength during an application, without sending a shock through the patient whenever a plug is moved. For this purpose a double rheophore, that is to say a rheophore bifurcated at one extremity may be used, with each of its branches carrying a plug. The plug actually in use should never be removed before the second has been fixed to the hole above or below, as the case may be. A perfectly smooth graduation of the current is thus obtained.

With a little dexterity, however, one may accomplish this end without mechanical contrivances. It is sufficient, the full number of cells wanted being included between the rheophores, to apply one of the electrodes very gradually to the skin, or over a thick part of it, and gradually draw it to the desired spot. Thus when the face has to be galvanised one of the electrodes being fixed to the neck the other is applied to the scalp and gradually drawn towards the part of the face upon which it is intended to rest. The opposite manœuvre enables us to remove the current without sudden break.

The Sledge Collector.

This collector is exemplified in the familiar Stöhrer's zinc-carbon battery. A sledge running in a groove carries on its inferior surface two springs which come successively into contact with metallic pieces arranged in two rows. To these pieces the wires from the cells are attached. The sledge carries on its upper surface two binding screws, connected with the springs for the rheophores, and a commutator. The cells are thus taken two by two; but by a simple arrangement, which consists in making one of the springs moveable, Baur has improved the sledge, and made it possible to take the elements one by one.

The springs are made long enough to avoid breaks in the current when the sledge is moved along, exactly as in the dial collector.

Fig. 59.

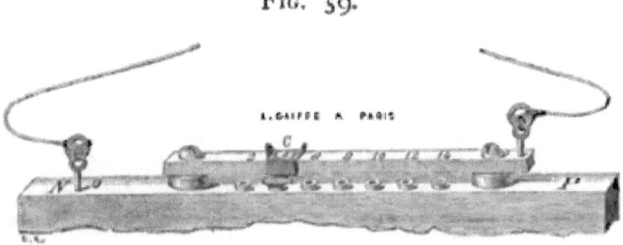

Fig. 59 shows a simple sledge collector which may be adapted to batteries such as the one shown in fig. 32. A stiff metallic bar is screwed so as to allow a sliding piece C to come into contact with the studs (numbered on the bar 2—14). One of the rheophores is fixed to the first stud O, connected with the first zinc (compare fig. 26); whilst the other electrode, P, is connected through the metal bar and slider with one of the studs, and thereby with the carbon of the cell the number of which is indicated on the bar.

The Dial Collector.

This is the most generally adopted form of collector. In the centre stands a pivot upon which a metallic spring or switch revolves like the hand of a watch. (M, M', in the two dials, fig. 60) The free extremity of the spring comes successively into contact with a number of metallic pieces, screw heads or studs, (0, 2, 4, etc.,) arranged in a circle around the central pivot. The negative pole of the first cell is directly connected with one of the binding screws R R' to which the rheophores are attached. The positive pole of each successive cell, (or 2nd, 3rd, etc. cell) is connected in order with the metallic pieces of the dial, and through the spring and pivot to the second binding screw. The dial studs are numbered so as to indicate the number of cells included in the circuit. The free extremity of the spring is made broad enough to come into

Fig. 60.

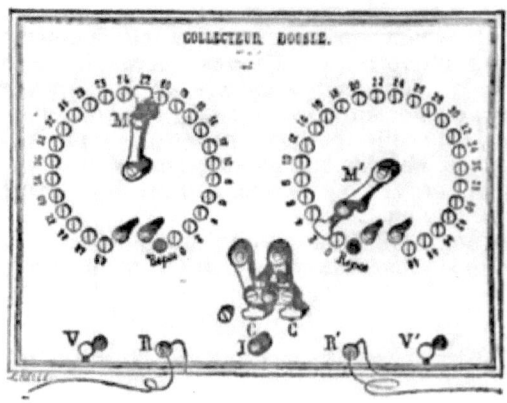

The **double collector** consists of two dials, each with the same number of studs or screws, which correspond to the number of cells. The negative pole of the first cell is connected with O, the positive of the 48th with 48, and the wires connecting the second to the third, the fourth to the fifth cell, and so on (see figs. 26 and 61) in which however the cells are taken singly, not by twos, are attached to the screws, marked 2, 4, etc. respectively. The screws in both dials having the same number are connected directly together by transverse wires.

M M' are switches revolving upon the studs and connected with the binding screws R and R' respectively. It is evident then, that the number of cells included between M and M' are thrown into action when the circuit is closed externally, and that the negative pole is on the side on which the handle rests upon the smaller number. When the two switches rest upon the same number, no cells are in circuit. Thus, in the figure, the cells in action would be 20 in number, from 3 to 22 inclusive. If the switch M' were pushed on to 20, only two cells, 21 and 22 would be active; if pushed on to 42, 20 cells would be again included, but M' would then be the positive pole. The current reverser C C' is described elsewhere, and the interruptor I, is a simple knob on a spring, which is pressed down with the finger. V V' are bolts for fastening the element board to the battery. Hinges not shown, fix it on the other side and allow of a free access to the interior of the battery.

contact with the next stud before leaving the previous one, so as to avoid breaking the current every time it is moved.°

The obvious disadvantage of the dial collector is that the first few cells of the battery become exhausted long before the last which are more rarely used. Among the various contrivances devised to remedy this drawback the simplest consists in using perforated studs, into any one of which the rheophore may thus be fixed. The current is taken from that point by means of the rotating handle. But the most perfect arrangement for the purpose is Gaiffe's double collector. Its advantages are that besides enabling us to distribute the work evenly over all the cells and to verify rapidly the state of the battery, immediately pitching upon the disabled cell if such a one exists, it allows us to reverse the current, the electrodes being applied to the body without sending a voltaic shock through the patient.

FIG. 61.

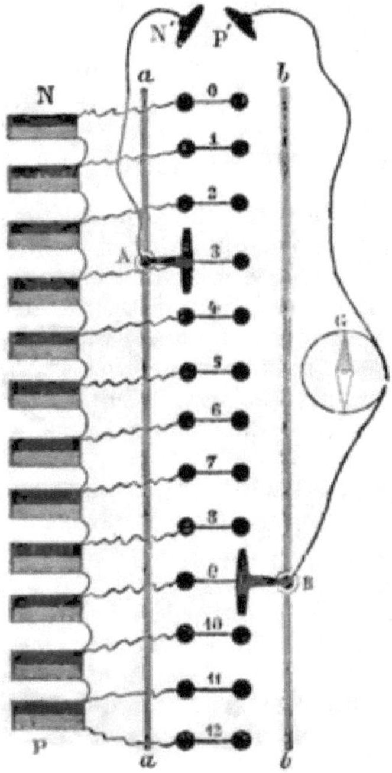

The appended diagram will assist the reader in realising the action of the double collector. Let 12 cells (in series) be connected as shown, to 12 pairs of studs connected by cross wires 0, 1, 2,—11,

* Great care must be taken *never* to leave the cell or spring in contact with two contiguous studs, as it short-circuits the included cells and rapidly runs them out. This recommendation applies to all collectors with smooth action. A glance at fig. 61 will show how this takes place.

12, the zinc of the 1st cells being attached to 0, the carbon of the 12th to 12. Two metallic pieces A, B, are made to slide along a a', b b', touching the successive studs of their respective rows as they move up and down. To A B the electrodes N' P' are attached. Now it is clear that the number of cells in circuit will always be that included between the two sliders (*e.g.* in the diagram 6 are in circuit viz 4 to 9 inclusive), and that the polarity of the electrodes depends on the relative position of the sliders (*e.g.* in the figure if A stood on 9, and B on 3, the polarity of N' and P' would be reversed: N' which is negative would become positive; and P' negative). The rule would be for a battery arranged as the one depicted that the electrode on the side of the smaller number (as indicated by the slider) would be negative.

Now imagine the studs to be arranged in 2 circles around A and B which would then be made to revolve on pivots, and you have the arrangement depicted in fig. 60.—The action of a simple collector may be understood by imagining the electrode P' to be connected with O, and the switch B and its row of studs to be absent.

(The wires connecting the cells to the studs are attached in this diagram not to the *middle* of the wire connecting cell to cell (as in fig. 26) but to one of its extremities; this is a matter of mere detail).

Rheophores.

By this name are designated the conducting cords which serve to attach the handle or plate electrode to the battery.

Nothing but actual experience can give an idea of the petty annoyances and loss of time inflicted by the use of bad rheophores. The stuff supplied by many instrument makers is perfectly useless, as the metallic threads inside are easily corroded or broken, and make very bad external contacts. On the whole thin telegraph wire, that is, copper wire encased in a sheath of India-rubber, is to be preferred, on account of its efficiency, cheapness, and durability. I have been much pleased with Gaiffe's improved rheophores, consisting of twelve fine copper wires enclosed in a double sheath of India-rubber and silk. They are more springy than the telegraph wire, and less apt to become entangled. It is convenient to have the positive and negative rheophores of different colour, or otherwise distinguishable; a knot, or other mark at each of the two extremities of one of the wires is sufficient to identify the polarity of the electrode to which it is attached without going each time through the process of following it up from one end to the other.

The drawer of the battery should always contain a few spare yards of telegraph wire, besides the two pieces—each about four feet long—actually used as rheophores.

For certain purposes it is necessary to have a double rheophore, that is one bifurcated so as to allow of two electrodes being attached simultaneously, as when it is desirable to diminish the local action of the negative pole; or to act upon two points at the same time; or to perform electro-diagnosis, as will be mentioned further on. A double rheophore is readily manufactured from two pieces of telegraph wire, about two and three feet in length respectively.

The longer piece is attached by one extremity to the battery, by the other to the middle point of the shorter, the two extremities of which carry the electrodes. The double rheophore mentioned with regard to the pin and hole collector may be made in this way, as well as the mutiple rheophores used in electrolysis.

Connections.

Speaking of rheophores I must advert to a very important practical point, upon which too much stress cannot be laid. Many failures in electrisation are due to the imperfect contact of rheophores with the pins, screws, etc. to which they are attached. Where a wire is simply tied to, or wound round, a piece of metal particular attention must be paid that the two metallic surfaces be in firm juxtaposition. Contacts are best established either by a screw or soldering. The bungling manner in which some makers connect the rheophores is simply disgraceful. Whenever air is admitted between two surfaces of metal conveying an electrical current, oxides are formed; moreover loose connections produce interruptions, hence disagreeable shocks to the patient. It is of course necessary that where screws and plugs are used, the surfaces of metal should be clean to begin with. This is best secured by a preliminary rub with emery paper, or scraping with a penknife.

Electrodes.

For purposes of local faradisation the *handles* should be about three inches long and one inch in diameter, and hollowed out about their middle so as to be conveniently held both in one hand, between the first and second, and third and fourth fingers respectively. For all other purposes (unipolar applications, labile applications to large surface, etc.), much larger handles may be used with advantage.

The familiar copper cups and sponges ought decidedly to be excluded from the battery. They are both dirty and unsatisfactory. If sponges are to be used, either plated, or vulcanite, or wooden cups should be provided so as to avoid the formation of copper salts, which impart a repulsive green discoloration to the sponge.

Far superior, however, to any sponge-holders are the carbon electrodes, consisting of a carbon disk enclosed in a cover of flannel or washleather, with or without the addition of a layer of fine sponge. This cover contains a more definite amount of water, hence its resistance does not vary as that of a sponge. Pressure may be made with the carbon electrode without drenching the patient with an overflow of water, and without the risk of touching the skin with the edge of the metallic cup which is very painful.

Fig. 62.

Interrupting handle. The disk screws on at V.B, attachment of rheophore; I, interrupting stud.

A clean cover may advantageously be slipped over the electrode every time it is used.º

Metallic disks are preferred by some and are quite satisfactory. Schoth has lately made for me disks of vulcanite coated with a thin metal plate, which combine lightness with solidity.

I cannot too strongly recommend the use of plates of lead and tin as electrodes; as regards both efficiency and convenience (when the electrode does not require to be moved about) their advantages are great. They are pliable and can be made to fit accurately any part of the body; they are readily slipped under the patient's dress, and are there kept in place by the pressure of the clothes, without any effort to hold them; or they may be fixed with a towel or turn of bandage.

Fig. 63.

Carbon disk electrode covered with wash-leather.

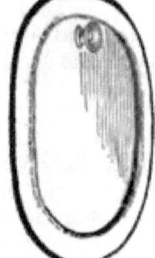

Fig. 64.

Like the disks they are to be placed in flannel or wash-leather covers,† and moistened with water. In order to avoid wetting the patient's clothes, one side of the cover is to be covered with oil-skin or india-rubber tissue. The plate is either permanently connected with the rheophore by means of soldering, or provided with an appropriate binding screw, (fig. 64).

The size or rather the surface, of the electrodes is a point of the highest importance,‡ but which hitherto has not been dealt with in electro-therapeutical treatises in the clear and definite manner it requires, except in the recent work of Prof. Erb. He gives the following list of standard electrodes which I adopt the more readily that the sizes he recommends correspond very closely with those I have been in the habit of using.

1. Fine electrode . . ½ centimetre (¼ inch diameter.)
2. Small electrode . . 2 cm. (¾ inch).
3. Medium electrode . . 5 cm. square, or 7·5 diameter, (2 inches, or 3 inches).
4. Large electrode . . 6 × 12 cm. (2½ × 5 inches).
5. Very large electrode . 8 × 16 cm., and more.

The "fine" electrode is used in electro-diagnosis for applying the

---

* The following statement, made on good authority, shows the necessity of attending to apparently trifling matters. Not long ago the child of a physician in Florence, was electrised at the hospital for some paralytical disorder. A few weeks after a distinctly syphilitic rash made its appearance. It was found on inquiry that the sponges used had been employed in the case of a patient infected with syphilis.

† Special care must be paid to these covers, which must be substantial, as otherwise unpleasant eschars will be produced at the edge of the plate.

‡ See page 44, and chapter on Treatment.

current to the motor points, small nerves or muscles. The "small" electrode is used in diagnosis for the excitation of the larger nerves and muscles; and in therapeutics for applications to the eye, etc. The "medium" electrode is useful for the examination of larger muscles, and for a large number of therapeutical applications (to the face, the neck, the limbs, etc.) The "large" electrode is used for galvanisation of the spinal cord, of large muscular masses, and deeper structures (*e.g.* sciatic nerve), of joints, etc. The "very large electrode" is used as "indifferent" electrode in the process of electro-diagnosis, and for the same purposes as the "large" electrode, when the patients are stout, or strong currents are required. It is indicated when the abdominal organs are to be influenced. I use it commonly in the galvano-faradisation of large muscular masses, etc.

Fig. 65.

Fine electrode.

A few words may be added here, with reference to the moistening of the electrodes. When the battery is weak, and especially when sponges are used, warm salt water is requisite. Otherwise pure water, cold or tepid, according to the season or sensitiveness of the patient is enough. The addition of a few drops of aromatic vinegar makes it pleasant, and enhances its conductivity. The prolonged and repeated application of salt to the skin irritates and roughens it; an objectionable result in the eyes of some patients, but which may be obviated by applying vaseline to the excited skin after each application. The addition of either salt or vinegar hastens the corrosion of copper, which, as it cannot be entirely banished from the construction of rheophores, should always be strongly plated. Nickel or silver plating ought, in fact, to be applied to all the accessories of a battery, dials, commutators, binding screws, etc.

Fig. 66.

Fig. 67.

Dry electrodes are used solely for the purpose of cutaneous faradization, and consist usually of a metallic brush or cylinder, without any covering. The brush is best made of nickelised copper wire, and has usually the shape of an ordinary painter's brush, as in fig. 66. For applications to large surfaces, it may conveniently be made in the shape of a clothes brush, with bipolar arrangement. The metallic cylinder, fig. 67, needs no special description. A metal or carbon disk, stripped of its cover, answers very well the purposes of a dry electrode.

Electrodes may be attached to either end of the handle, as shown in fig. 72, M and M'. For ordinary purposes it is best to have the connection as at M, that is at the junction of the handle with the carbon disk.

Many forms of special electrodes have been devised for the

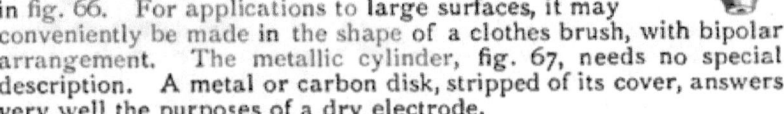

bladder, the rectum, the uterus, the throat, ear, eye, etc., and are required for special applications, (see figs. 68, etc.)

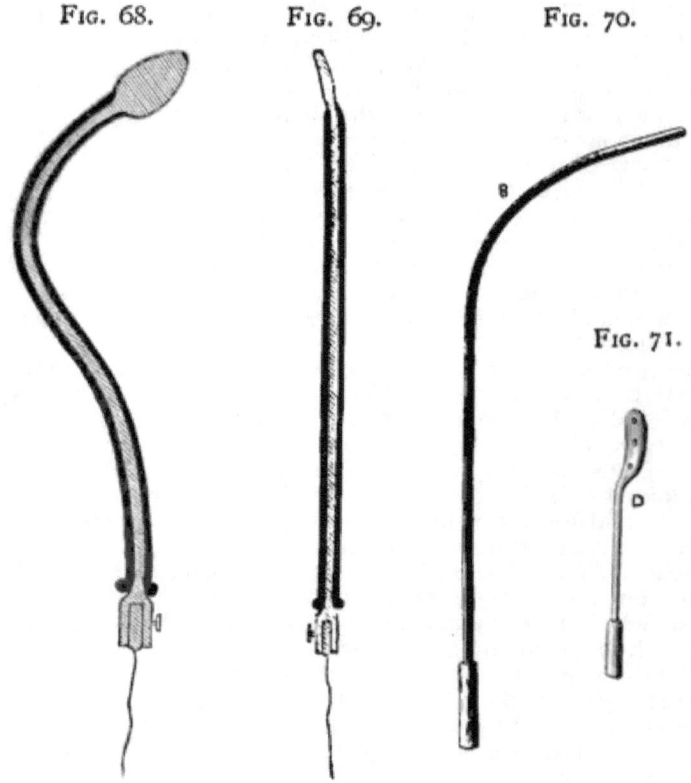

FIG. 68. FIG. 69. FIG. 70.

FIG. 71.

INTERRUPTORS.

THERE are two modes of interrupting the current during an application. First, by leaving one of the electrodes in contact with the skin, whilst the other is successively applied and removed. Second, by leaving both *in situ*, and making and breaking the current in the metallic portion of the circuit. The former is sufficient for rough diagnosis and treatment; but for accurate work a contrivance for effecting the latter is necessary.

In some batteries a special interruptor is provided, or the commutator is used for the purpose. But this arrangement is far less convenient than the interrupting handle, which is indispensable for

INTERRUPTORS. 93

electro-diagnosis. It consists of an ordinary electrode handle, inside which a spring is concealed, (fig. 62). Pressure being made with the finger upon an ivory stud protruding externally, the spring is depressed and contact broken. The contact surfaces are to be, of course, kept clean, and are best made of platinum. Other shapes of interrupting handles have been devised, such as the one represented in fig. 72.

Pedal interruptors have also been devised, but are mainly used for producing slowly interrupted faradic currents.

Toothed wheels for obtaining rapid interruptions are provided in some batteries. The objection to them is that they often give double shocks, and are very liable to get out of order.

Automatic interruptors, as the toothed wheel moved by clock-work, the metronome and mercury cups, etc., have hitherto been rarely used for medical purposes. Brenner's interruptor is an ingenious application of an electrical contrivance adopted in some telegraphs. Dr. Onimus' interrupting apparatus for producing numbered and measured interruptions to an electrical current is represented in the following wood-cuts.

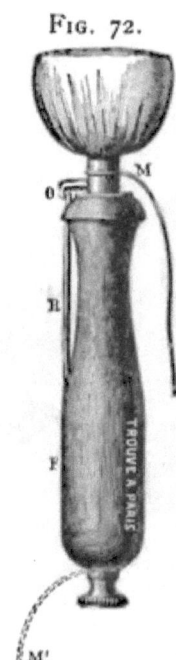

FIG. 72.

Interrupting handle and carbon disk.

FIG. 73.

Fig. 74.

This apparatus has been constructed for Dr. Onimus by Trouvé, of Paris.

The first figure shows the interrupting part by itself; the second shows in addition the inductorium.

Binding screws 1, 2, are for attaching the wires for obtaining the interrupted galvanic current. 3, 4, for the wires of the battery for the faradic current. 4, 5, for the rheophores when the primary or extra-current is required; 6, 7, for the secondary; 5, 7, for both primary and secondary.

The interruptor consists of a cylinder E, made up of 20 disks carrying from 1 to 20 pins, in regular progression. A clock-work arrangement causes the cylinder to revolve.

A fly wheel J, regulates the number of revolutions of the cylinder (from 1 to 4 a second).

A lever H, with mercury cup arrangement, is made to assume any position along the cylinder by means of the rod K. This lever every time it comes into contact with a pin, breaks the current. In this way from 1 to 80 interruptions a second are obtainable.

M primary coil, C metallic tube to regulate the extra-current; B, B', secondary coils of different length and thickness; D sledge for graduating the secondary currents; L is for winding up the clockwork. I and G, represent the lever for starting and stopping the clockwork, in its respective positions.

An ingenious contrivance not shown in the figure is added to the interruptor, by means of which the successive passages of the battery current are made of strictly equal duration.

CURRENT REVERSER OR COMMUTATOR.

IT is often necessary for purposes of diagnosis or treatment to be able to reverse the current through the electrodes and the body.

For this purpose several contrivances have been devised. One of the most familiar is the barrel commutator and its modifications, adopted by Stöhrer, but need not be described in detail. The only observations to be made here are, that the handle of the commutator should be large enough to allow of its being worked comfortably; and that when the full effects due to "voltaic alternatives" are desired, the commutation should be immediate, *i.e.*, that there should not be an interval during which the current is opened in the act of commutation.

Another convenient commutator often seen in French and English batteries, is made in the shape of two springs playing upon three studs and similar to those used for the dial collector (fig. 75).

FIG. 75.

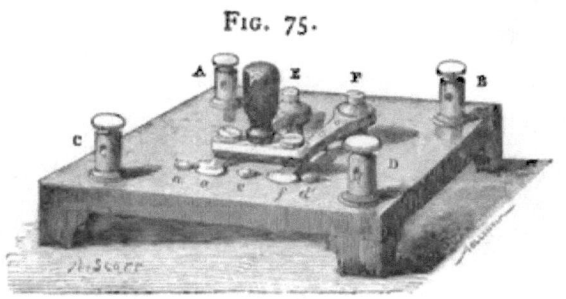

Current reverser, or commutator. A B, binding screws receiving the rheophores from the battery, and connected with EF respectively, which are the pivots upon which the two switches E e, F f, revolve. The switches consist of a spring, e, f, fixed to the inferior surface of a rigid piece of brass pivoted as just mentioned; a transverse piece of ivory, surmounted by a handle connects the two rigid pieces, and serves to move them laterally.

C D, are the binding screws for attaching the external rheophores, C being connected with c, D with d and d'. The latter are metallic studs upon which e and f are made alternately to rest. A glance at the picture will show that the direction of the current is changed through C D, according as the position of e and f, is on d and c, or on c and d' respectively. (See A, B, C, in fig. 77).

The see-saw mercury commutator may be used for stable batteries; but has hitherto been mainly confined to the physiological laboratory.

CURRENT ALTERNATOR, REVERSER, AND COMBINER.

The apparatus (fig. 76, 77) is made up of two reversers of the kind just described fixed back to back on a tablet of wood. The one to the right (1, 2, 3 in the Diagram) differs from the other in having the two external studs (1, 3) prolonged into metallic bands so that when in the position depicted, the springs rest upon their inner extremities. Two pairs of screws G, F, receive connecting wires from the poles of the galvanic and faradic apparatus respectively. When the springs rest upon 1 and 2 the galvanic current alone circulates; when upon 2, and 3 the faradic. When upon 1 and 3 (as in the diagram) the galvanic current passes through 2 to F−; thence through the coil to F+, to 3, and finally reaches ± or ∓ (to which the electrodes are fixed) according to the position of

FIG. 76.

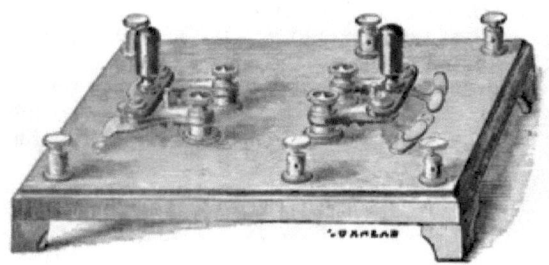

FIG. 77.

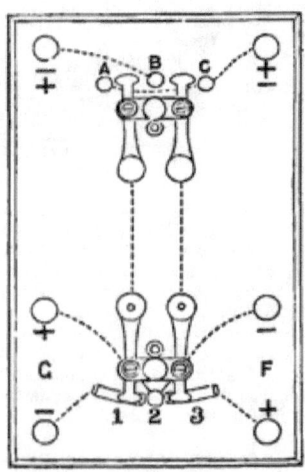

the reverser. Having traversed the body it completes its circuit through the opposite half of the apparatus, G and the battery. If the faradic current is flowing at the same time it accomplishes the same circuit. When all connections are established as in the diagram, the galvanic polarity of the electrodes attached to ∓ and ± is easily determined if we remember that the commutator A B C always points towards the positive terminal.

The apparatus serves the purposes of a current *alternator*, that is enables us to send either the galvanic or the faradic current through the electrodes; of a current *combiner*, that is, enables us to send both currents together through the body; and of a current *reverser*, that is enables us to change the direction through the body either of the galvanic or faradic current flowing singly, or of both currents flowing together.

The advantages of the instrument are obvious and manifold. It may either be made as a separate apparatus for the electrical table, or fixed to the element board of an ordinary galvanic battery which offers then the advantages of a "combined" faradic and galvanic apparatus without its drawbacks—especially if its drawer be fitted with an induction coil. A "separable combined" battery may be thus obtained. The comfort of being able to pass immediately from the galvanic to the faradic current without removing the electrodes is very great, in diagnosis especially; whilst the process of "galvano-faradisation," which the apparatus enables us to carry out without trouble, is one of great therapeutical value.

Chapter II.

ELECTROPHYSIOLOGY.

The task of electrophysiology is double. It includes both the study of living tissues as sources of electricity; and the study of electricity as a means of influencing the functions and of modifying the excitability of those tissues. The discovery of the constant current, and the early history of this new manifestation of electrical force (Galvanism or Voltaism, as it is called after the names of the two great men who were the pioneers of this field of research) are intimately bound up with the first of those electrophysiological problems, the question of "animal electricity." It is however with the second only that we are concerned, and mainly in as far as it can be studied on the living human organism. It will therefore be my aim to present the leading phenomena of human electrophysiology in a clear and practical manner dwelling chiefly upon the effects of electricity as observed on motor nerves and muscles which offer the greatest facilities for experiment.

A. Motor Nerves.

It was long ago observed that when a current of electricity from one galvanic cell was applied to the sciatic nerve of a recently killed frog, the leg was jerked both at the time when the electrodes were made to touch the nerve, and also when they were removed. During the time the current was flowing through the nerve the muscles were quiescent.

It was also noticed that the amplitude of the muscular contractions diminished and increased (within certain limits) with each diminution and augmentation of the current-strength used. Du Bois-Reymond, however, showed that there was another factor which determined the amount of reaction of a nerve to the electrical stimulus, besides the absolute strength of (that is the quantity of electricity conveyed to the nerve by) the current. This factor is the rapidity with which that quantity reaches its maximum in the nerve.

If a current of, say, 1 milliampere, be very gradually made through a nerve no contraction follows; whereas the effect is marked if a current 10 or even 100 times weaker is made suddenly. It is therefore the rapidity with which the electrical density changes in the nerve, as much as the absolute value of that density, which excites.

In all the experiments described below it is assumed that a galvanometer (or galvanoscope) is included in the circuit, and that the observer is alive to the fallacies that would be introduced into the results if the changes in the current strength, due to diminished resistance in the body, were not eliminated by the method described above (page 39) or compensated by a corresponding diminution in the number of cells used. Once the skin thoroughly soaked the dangers arising from that source are much diminished.

EXPERIMENT I.—Take a galvanic battery in good working order, and connect two electrodes with its poles. One of these electrodes, of large size, is attached to the positive pole* and fixed securely on a distant part of the body (sitting upon it answers very well); the other, of small size (not more than one inch in diameter) is applied immoveably and with uniform pressure over a superficial nerve such as the peroneal at the head of the fibula, or the ulnar at the elbow or at the wrist; a small metallic plate, as described in chap. I and fixed by a turn of some elastic bandage, answers best the purpose. Now by means of the collector gradually put into circuit 1, 2, 3, 4, etc., up to 12 or 15 cells. As the current increases no motor effect is produced. The nerve is not excited, the change of electrical density not being rapid enough to act as a stimulus.

EXPERIMENT II.—Without altering anything in the battery or electrode, break† the circuit by means of the interruptor or by disconnecting one of the rheophores, and make it again by bringing the two metallic surfaces together. A lively contraction is produced. The current has almost instantaneously reached its full strength when the circuit was restored; *i.e.*, the change of density in the nerve has been almost instantaneous.

The experiment may be varied in the following manner. Leave the large electrode applied to a distant part of the body, put 15 cells in circuit, and apply very gradually the other electrode (thoroughly moistened as it always must be) over the ulnar nerve, or apply it to the palm of the hand and draw it firmly but slowly up to the wrist. No contraction ensues. The current in the nerve

---

* We shall throughout designate the electrodes as *anode* and *kathode* according as they are attached to the positive and to the negative pole of the battery respectively.

† The terms make and break, closure and opening are used to designate the acts by which the current is caused to flow through the body or other conductor (make or closure) and caused to cease (break or opening).

has in both cases gradually been made. Now lift the electrode and apply it rapidly to the nerve: a contraction occurs.*

EXPERIMENT III.—The same series of phenomena (contractions proportional to the suddenness of the electrical change) are to be observed with reference to the gradual and sudden *break* of the current. Only for this purpose you must (for reasons explained below) apply the anode to the nerve, and use stronger currents.

It would, however, be incorrect to conclude from these experiments that interruptions of the current are absolutely necessary to produce contractions.

Du Bois-Reymond's law applies chiefly to the frog's nerve, and holds in the case of the human nerve only for currents of moderate strength. Beyond a certain point the continuous and even flow of a current stimulates the nerve and gives rise to contractions:

EXPERIMENT IV.—Repeat Exp. I., taking care to place the kathode on the nerve. Go on increasing the current until you see that the muscles contract. They will remain shortened for a variable period during which the galvanometer shows that no alteration takes place in the current-strength. (Some persons and some points over certain nerves are better suited for this experiment than others). It may be varied by suddenly making a strong current.

If you try each of the experiments III. and IV., applying both poles successively to the nerve, you will find that you do not get equally good results, according to the one you place on the nerve. In fact you often fail to obtain with the one what you obtain readily enough with the other. The natural conclusion is that as the two poles differ in their chemical, so they differ in their physiological, properties.

EXPERIMENT V.—Having disposed the apparatus as for Exp. I., and taking care that the epidermis be thoroughly soaked before beginning, find out how many cells (or better, how many milliamperes) are required to produce a minimal (*i.e.*, the first visible) contraction by closing and opening (at uniform intervals) the circuit, 1st, with the kathode, 2nd, with the anode on the nerve. *Repeat the experiment three or four times.* The means of the numbers so obtained show you that 1st, the earliest contraction is obtained when the circuit is closed the kathode lying over the nerve (kathodic closure contraction, KCC);† 2nd, the next in order is when

---

* The observer will do well to accustom his eye to the mode of reaction of healthy nerve to a simple excitation as judged by the muscular contraction. The latter runs a definite course in which we may distinguish the period of ascent, (or of contraction proper), that of contracture (duration of maximum shortening), that of descent (or of relaxation). The myographic curve obtained on a rotating cylinder registers on a magnified scale the three periods. In a healthy muscle the ascent is abrupt, the summit sharply rounded, not protracted, the descent rapid.

† I spell "kathodic" with a K, because it is more convenient to represent the negative pole by a K in the formula. It allows to use the letters KC to mean a kathodic closure, and CC to mean a closure contraction. Again kathodic and and anodic should certainly be used instead of the mongrel, barbarous, and ungrammatical "cathodal," "anodal," which have crept into English electrotherapeutical literature. We say "methodic" not "methodal," "periodic" not "periodal." Strictly, anode and kathode should be spelt and pronounced anod and kathod, but the error is of too long standing to justify an attempt at reform.

the circuit is closed, the anode lying over the nerve (anodic closure contraction, ACC); 3rd, closely upon the heels of ACC, or along with it, or even sometimes before it, appears the anodic opening contraction (AOC), due to opening the circuit when the anode is over the nerve. Finally, with strong currents, the kathodic opening contraction (KOC) is obtained.

You will notice that (within certain physiological limits) the amplitude of each contraction is increased every time the strength of the current necessary to obtain that contraction is increased. Thus KCC will be strong when you obtain it with the current necessary to excite an ACC or AOC; very strong with that necessary to excite a KOC. ACC and AOC, will be strong with the latter current.

We have already found that the flow of a strong current (its Duration, D) produced a persistent contraction (Tetanus, T), when the negative, or even the positive, pole rested over the nerve. We thus have to consider at which stage the apparition of the kathodic and of the anodic duration tetanus (KDT and ADT) occurs with reference to the chronological sequence of the closure and opening contractions. On repeating the experiment and watching for the first signs of KDT, you will find that it usually precedes the occurrence of KOC. The ADT is usually difficult to bring out with bearable current strengths.

In all these experiments special care is to be taken not to mistake the action of other muscles, through which the current diffuses itself, for contractions of those supplied by the ulnar or other explored nerve below the excited point, and to which alone the preceding statements are applicable.

We may tabulate the phenomena observed as follows—indicating the increased energy of the contractions by the symbols $CC'C''$.

The letter D need not be used; T following KC and AC indicates that the closure contraction does not subside at once, but is protracted during the time of flow.

1. Very weak current . . . KCC.
2 and 3. Weak currents . . KCC', ACC and AOC.
4. Moderate current . . . KCT, ACC' and AOC'.
5. Strong current . . . KCT', ACC'' and AOC'', KOC.
6. Very strong current . . KCT'', ACT and AOC''', KOC'.

For reasons to be stated further on, numerous exceptions will be found to occur in the order and intensity of the minor members of the series as here given, but the following formula must be remembered as one of universal application to healthy human nerves:—

1. Weak current . . . KCC.
2. Medium . . . ACC, AOC,
3. Strong . . . KOC.

EXPERIMENT VI. Repeat experiment V at different points along the ulnar, the peroneal and various other superficial nerves. You will find that whilst KCC is invariably the first in order, and KOC the last, ACC and AOC vary considerably with relation to the strength of current necessary to produce them.

EXPERIMENT VII. Place the small electrode on the median at the elbow (inner side of biceps tendon) and alternately make it anodic and kathodic—AC and KC. Use in each case a current strength sufficient to bring about equally well marked contractions. Observe exactly the movement of the arm at each event. KC is a more powerful stimulus than AC, but you will find also that its effect is not quite the same and that different muscles contract at AC and at KC. This must mean that the excitations at KC and of AC do not fall upon the same nerve fibres in the two cases. Under favourable conditions you may also perceive that the same movement is produced by AC as by KC. It is difficult to demonstrate that the KC and AC are followed by an excitation of the same parts; but this may occasionally be done. The reason of these phenomena, we shall presently show, is to be found in the diffusion of the current in the nerve and its surroundings.

EXPERIMENT VIII. Place the anode over the nerve and take a current strong enough to produce a very weak AOC, when the circuit is opened after the current has passed for one or two seconds. Now keep the circuit closed for one or two minutes: on opening it the contraction will be found much increased. Again a current just too weak to produce an AOC after one second's duration will call it forth if allowed to flow a little longer, though the skin has been thoroughly soaked and exposed to the current before. The apparition of AOC is determined therefore not only by the strength of the current, but also by the time the current has flowed through the nerve, before the circuit is broken.

EXPERIMENT IX. Take a strength of current that gives a distinct AOC after five seconds duration. Go on breaking the current every five seconds, closing it again immediately. After a variable number of repetitions, the AOC will gradually diminish and finally disappear. The ACC on the other hand suffers no change.—KOC may under similar conditions be seen to dwindle and disappear; but the demonstration of this fact involves a very painful process.

The opening contractions are diminished by repetition. The explanation of this phenomenon is still obscure.

What we have hitherto learnt from our experiments is mainly that the make and break contractions at the anode and kathode, occur successively and in a certain order, at different stages of increasing current strength; at the same time we have observed significant differences between the circumstances attending the various members of the series of contractions.

It has been for a long time known that the passage of a galvanic current through an exsected frog's nerve, altered in peculiar ways the properties of that nerve, which is described as being then in the state of "electrotonus." It is found especially that the excitability of that nerve is modified; at the point of application of the anode its excitability is diminished (anelectrotonus); at the point of application of the kathode its excitability is increased (katelectrotonus).

Now since the classical researches of Pflüger—embodied in a

work which stands as a model of experimental inquiry—the phenomena of electrical nerve-excita*tion* are all reducible to the fact of electrotonus, *i.e.*, of modified excita*bility*. A nerve is excited when it passes from a condition of lower to a condition of higher excitability; that is when it passes from the normal state to the state of katelectrotonus or from the state anelectrotonus to the normal state. In other words a nerve is excited by the appearance of katelectrotonus, and by the disappearance of anelectrotonus. Replace the terms kat- and an-electrotonus by those of kathodic and anodic influence; and those of appearance and disappearance, by those of make and break. We thus arrive to the general statement that a nerve is excited at the kathode, when the current is made (KCC); at the anode, when the current is broken (AOC). In other words there occurs a closure contraction CC, originating at the point *only* where K is on the nerve: an opening contraction OC, originating at the point *only* where A is on the nerve.

Such being the results of the closest investigation of the facts, as observed on the exsected frog's nerve, the reader will at once ask, how it is that the phenomena displayed in our experiments on the human nerve stand in such absolute contradiction to them. Here we have not only a KCC and an AOC—but also a KOC and an ACC—that is an opening contraction with the kathode only over the nerve; a closure contraction with the anode only over the nerve.

When a law is so well established as that of Pflüger, and we find that facts so well observed and controlled as the polar reactions of human nerves appear to militate against it, we may safely suspect some difference in the conditions under which the two sets of experiments have been conducted. In this case we have not far to go: from the very statement of the case it appears that the electrodes are placed *on* the frog's nerve, in direct contact with it; *over* the human nerve, not in direct contact with it. Moreover we have said that the nerve is exsected, isolated, in the one case; whilst it necessarily remains embedded among other tissues in the other. Finally both electrodes are usually placed on the frog's nerve, one electrode only over the human nerve.

Let us consider how far these different conditions explain the different results of the experimentation in the two cases.

When to an isolated nerve two exactly similar electrodes EE' are applied, the whole current enters at the point of application of the one and makes its exit at the point of application of the other: the density of the current in the nerve is uniform and the polar influences (kat- and an-electrotonus) are distributed through its whole diameter, in the neighbourhood of the kathode and the anode respectively. A certain length of the nerve is under kathodic, another length under anodic, influence; between the two is a neutral zone where the excitability is unaltered. These facts will be found fully described and illustrated in the usual texts-books of physiology and need only be mentioned here.

Now when the human nerve is placed under the influence of a

MOTOR NERVES. 103

pole applied to the skin over it, the distribution of the kathodic and anodic influence is very different. The nerve being no longer isolated, but surrounded with various tissues, the bulk of which (muscles) are better conductors, and of much larger diameter, than itself; the laws of derived (or diffused) currents make it at once plain, that no such localisation of the current is possible as exists in the exsected nerve. Diagram 78 shows what occurs. Directly under the electrode the current in the nerve obtains its greatest density: this region (marked + + +) we may conveniently term the

FIG. 78.

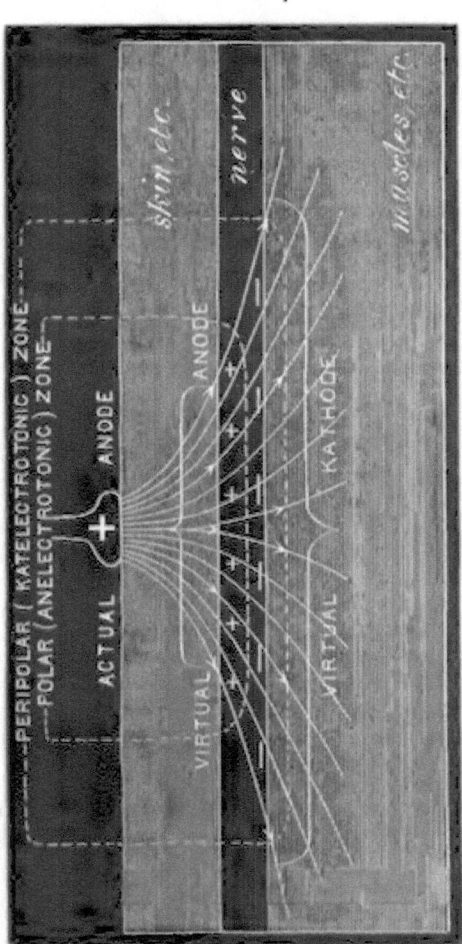

Diagram to illustrate the points of entrance and exit of the current in a nerve submitted to percutaneous excitation with one electrode, and the consequent formation in it of two zones of opposite electrical character; anodic and kathodic.

polar zone, and is necessarily of the same sign as the actual or apparent electrode;* (kathodic if the latter be the kathode, anodic

FIG. 79.

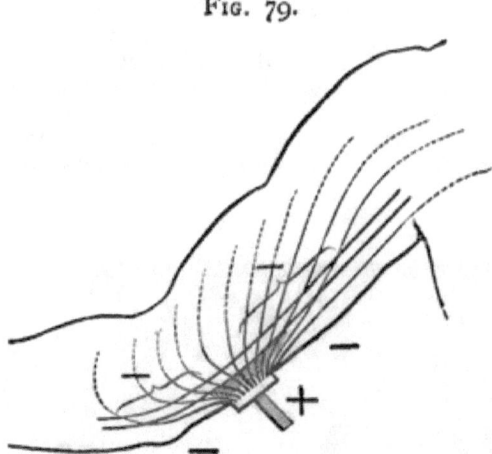

This diagram represents the same phenomena as the former. The anode is supposed to lie over the ulnar nerve, whilst the kathode rests on the trunk. The polar anodic zone is shaded. The two brackets show that owing to the diffusion of current the unipolar excitation of the imbedded nerve sets up in it a descending as well as an ascending current. The signs — indicate that under such circumstances the electrotonic condition of the nerve in the neighbourhood of the electrode may be found the opposite of that set up immediately under the electrode.

if it be the anode). Suppose E to be anodic: owing to the considerable diffusion of the current from the nerve into the muscles, etc., the nerve at a very short distance of the electrode is practically devoid of current. The current which has entered it in the polar zone P, has left it in a *peripolar* zone, — — —. Now, from the definition of the terms anode and kathode, (*i.e.* points of entrance and points of exit of a current into and from an electrolyte) as the polar zone was anodic so the peripolar zone is kathodic. By placing the actual electrode on the skin *over* the nerve, we have therefore created two *virtual* electrodes *on* the nerve, and which may be defined as the portions of the tissues surrounding the nerve from which and into which the current passes in its course through the nerve. They are represented in diagram 78 by the two brackets.

* It may be observed here that we say nothing of the *direction* of currents in the nerve. The phenomena of electrotonus are now generally recognised as purely polar effects; and by speaking of ascending or descending currents, physiologists merely intend to express the relative position of the two poles relatively to the centre and periphery. When experimenting on the human nerve with the "unipolar" method here described and used, it is obvious that no "direction" of current can be said to exist in the nerve, since the only portion of it which is traversed by currents of effectual density, is seen in the diagram to be so in opposite direction. Moreover the fact that the position of the indifferent electrode (above below or opposite the exciting electrode) does not affect the results, shows conclusively that the direction of currents plays no part in the results obtained.

These considerations give us the solution of the apparent contradiction between the phenomena observed on the frog's and on man's nerve. In both cases there is both an anode and a kathode on the nerve, and consequently both an opening and a closure excitation,[*] whichever electrode be applied externally. But if the analogy is fundamental between the two cases, these offer minor points of difference which are of the highest moment for the full explanation of the facts. 1st. In the frog experiment, the kathodic and anodic influences arise in zones of equal current density. 2nd. These influences are localised through the whole diameter of the nerve. In the human nerve the opposite conditions prevail. The peripolar zone is of larger area than the polar; the current in the former is less dense; therefore the excitation in it is, cæteris paribus, less energetic. The peripolar zone in the second place tends necessarily to prevail on the aspect of the nerve away from the skin, whereas the polar zone tends to be localised in the superficial aspect of the nerve.

The latter fact gives us a key to the results obtained in experiment VII. Different nerve fibres are stimulated by closures of the current when the actual electrode is alternately kathodic and anodic; in the former case the excitation falls in the polar, in the latter in the peripolar zone; each of these regions being kathodic in turn. The AO and KC excitations both fall in the polar zone, hence excite the same part of the nerve; hence the similarity of the muscular effect produced in the two cases.—The former fact explains how it is that KCC always precedes ACC; kathodic closure excites in the polar, anodic closure in the peripolar zone: the current in the latter being less dense than in the former, a higher strength of current is required to obtain the effect.—Experiment III has shown that ACC and AOC vary in their order of apparition. We are now able to explain the reason of this fact. Experiments on the exsected nerve teach us that appearing katelectrotonus is, *per se*, a stronger stimulus than disappearing anelectrotonus; and we know that the denser the current the more effective it is as an excitant. If a current of one ampere flows through two nerves containing 1000 and 2000 nerve fibres respectively, the current is twice as dense in the first instance than in the second. Each nerve fibre conveys 1 m.a. of electricity in the first, .5 m.a. in the second. In other words the current is twice as strong in the former nerve fibres than in the latter, and excite them in a proportionally higher degree. Now the relative area of the polar and peripolar zones, depends entirely upon physical conditions, which vary in different experiments made on the same subject; (relative position of electrode, and conductivity of tissues surrounding nerve). Therefore ACC will be produced by the same current

---

[*] It may not be superfluous to state expressly that the symbols KOC, ACC, etc., refer simply to what happens when the material electrode, kathodic or anodic, being placed on the skin, the circuit is closed or opened. They must not be taken to imply that under those conditions a closure contraction can be of anodic origin, or an opening contraction of kathodic origin—but to indicate the polarity of the actual electrode.

strength as AOC (after a given AD); when the greater density in the polar zone, exactly neutralises the greater energy of the appearing katelectrotonus in the peripolar zone: in other words, taking the energy of appearing katelectrotonus as $2x$, that of disappearing anelectrotonus as x, when the relative densities in the zones where these changes occur are as $2z$ to z (or reciprocally when the area of the anodic polar zone is one half of that of the kathodic peripolar zone), for then $x \times 2z$ in anodic zone $= 2x \times z$ in kathodic zone. Likewise ACC will be greater or less than AOC, (appear sooner or later, with greater or less current strength), according as the density in the kathodic peripolar zone is greater or less than that required to compensate the inferior energy of the vanishing anelectrotonus.

A moment's reflexion will show that KCC must always figure as the first member of the series, being due to the stronger stimulus acting on the denser zone; whilst for KOC the opposite obtains, both conditions being unfavourable to its elicitation; hence the great strength of current necessary to obtain this last member of the series.—Such are the facts by which the formula given at page 100 (1. KCC—2. ACC, AOC—3. KOC), is found to be but a manifestation of the so-called "laws" of contraction.

FIG. 80.

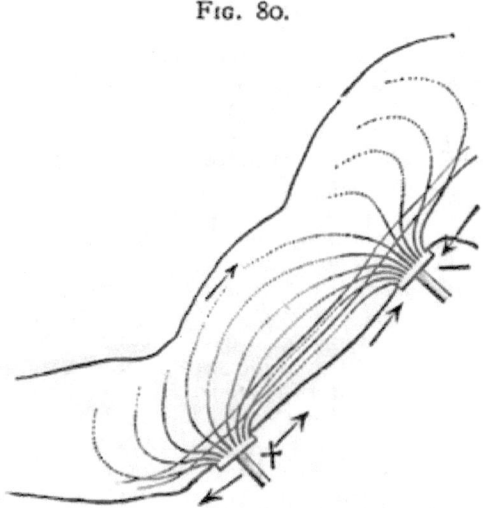

This diagram shows how with two electrodes on the nerve, the direction of the current in it is not single, owing to the diffusion of current. It also shows how four virtual electrodes then exist on the nerve. The dotted lines show the regions where the density of the current (*i.e.*, its physiological action) is least (cf. remarks on the "polar method" in the chapter on Treatment).

As a corollary of the previous discussion, it becomes apparent that nothing is gained by placing the two electrodes on the skin along the course of the embedded nerve. At every make the nerve would be excited at two points, at every break at two other points: the nerve would have four virtual electrodes applied to it. The conditions of the experiment would be complicated beyond all means of reckoning. Hence the impossibility of arguing from results so obtained. A curious superstition still prevails however in some quarters, respecting a supposed influence of *the direction* of the current in the nerve. The current it is said acts differently when it flows in the same direction as the normal nerve impulses (motor or sensory), or in the opposite direction.* Now 1st, all the phenomena observed in the frog's nerve receive a simpler explanation on the "polar" hypothesis. 2nd, on man (fig. 79) the portion of the nerve in which the electrical density is greatest, *i.e.*, *under* the electrode, is pervaded by currents in both directions, whilst the part of the nerve *between* the electrodes is practically devoid of current, at least with bearable strengths; for taking the diameter of the nerve and its conductivity as equal to one each, the conductivity of the surrounding tissues would then be equal to about five, and their diameter to ten, twenty or more, according to the part selected. Hence the actual quantity of electricity travelling along the nerve, in the intra-polar region (taken at 12 or 15 inches of length) would be as 1×1 to 5×10, or to 5×20, (that is amount to $\frac{1}{50}$th, $\frac{1}{100}$th, or less, of the whole current), as the case might be.

The object of having one of the electrodes on a distant part of the body, during electrical investigation of the embedded nerve, is to avoid having four poles on the nerve instead of only two as is the case in physiological experiments. Every conductor in the process of electrisation must have two poles at its surface, in other words a series of points of entrance and a series of points of exit for the current;† in the case of the embedded nerve, the areas occupied by those two series of points (the virtual electrodes) differ in extent. Now since KCC and ACC are both due to the appearing katelectrotonus in the polar and peripolar zones respectively, the difference between the currents necessary in every case to produce these contractions, must be proportional to the difference between those areas; for it requires a current twice as strong to obtain the same density in the area $2x$ as in the area x. We thus have in the strengths of current required to obtain a KCC and an ACC, a means of estimating the rate of current diffusion, *i.e.*, the relative extent of the polar and peripolar zones, in nerves. The area of the virtual electrodes is directly proportional to the strength of the currents necessary to obtain a (minimal) KCC and a (minimal) ACC.

EXPERIMENT X.—Dispose the electrodes as in Exp. 1., and take a number of cells sufficient to produce a KCC and an ACC. Estimate the amplitude of the muscular contractions obtained by a series of simple kathodic closures and another of simple anodic

---

* In all our experiments it matters not where the "indifferent" electrode lies with reference to the "testing" electrode, above, below, or opposite.

† This apparently simple and elementary proposition has too often been overlooked by the adherents of the "unipolar method" in electrotherapeutics, and consequently a vast mass of erroneous theories are to be found in many of their writings.

closures, and compare it with the amplitude of corresponding series of KCC and ACC, made by means of the "commutator." You will find that the effect of "voltaic alternatives," as the sudden reversal of the polarity of the electrodes is designated, is much more energetic than that of simple closures.

EXPERIMENT XI.—Compare the effect of reversing the current after short and after long intervals of flow. You will find that the effect is greater when the electrode on the nerve becomes kathode after it has been anode (and *vice versa*) for several seconds than when it has been so for only the fraction of a second. Compare also the effects produced by reversing the current very suddenly with those obtained by allowing the circuit to remain broken during intervals extending from the fraction of a second to one second and more, in the passage from one pole to the other. You will find that, *cæteris paribus*, the effect is the more marked, the less has been the interval elapsed in effecting the change of polarity of the electrode.

In order to understand the nature of these phenomena let us return to the diagram (fig. 78). When the electrode on the nerve is alternately changed from anode to kathode and from kathode to anode, a series of closure excitations are given, which fall alternately in the polar (when the electrode becomes kathodic) and the peripolar (when the electrode becomes anodic) region respectively. Now in either case the excited region had just before been under anodic influence, and physiology teaches us that as we shall presently demonstrate on the human nerve, the instant the polarising current ceases to flow the anodic region passes into a state of increased excitability. This augmentation is the more marked the longer the anodic influence has lasted. We see therefore how it is that voltaic alternatives act more powerfully than simple closures of the circuit, and that their action is intensified by previous current duration. We understand also why rapid reversals are the more effectual; for the positive modification after anelectrotonus diminishes rapidly after the circuit has been broken; the longer the interval which elapses between the polar change of the electrode the less the hyper-excitability of the nerve will be, until it has returned to its normal state.*

EXPERIMENT XII.—Observe the effects of series of kathodic closure stimulations (with various current strengths) in which (1) the times of closure and opening of the circuit are equal; (2) the one long, the other very short. You will find (a myographic tracing is necessary, however, to discover these finer changes) that the effect varies with the rhythm. With short duration and long interval the contractions are equal; with long duration and very

* The addition of the opening and closure effects may, when the current is strong and the inversion rapid, explain the increase of an A to K contraction—but scarcely of a K to A contraction owing to the feeble effect of KO. Brenner's attempt to explain the effect of voltaic alternatives by the passage of the nerve "from + 6 to − 6, and from − 6 to + 6 current" is physically absurd, as well as disproved by direct experiment as I have shown elsewhere.

As mentioned at page 45, the current from a given number of cells is much stronger, when it is reversed through the body, owing to the polarisation of the tissues. See note at end of this chapter.

short interval they diminish; with long duration and moderate interval they increase. All these phenomena are due to the after effects of the katelectrotonic condition of the nerve. Experiments on the frog's nerve, and much more clearly still experiments on the human nerve, have shown that the portion of the nerve which is under kathodic influence passes when the current is broken into a condition of diminished excitability (negative modification) of short duration, and which passes then into a state of increased excitability (positive variation) of longer duration. There is diminution of the effect, therefore, when the stimulus falls during the period of negative modification; increase, when it falls during that of positive modification.

B. Faradic Excitation.

EXPERIMENT XIII.—Take an induction apparatus and connect the rheophores to the secondary coil. Place the electrodes as heretofore. Choose a convenient strength of current. Make and break the inducing current by moving the hammer with the finger, if the apparatus be not supplied with a proper contrivance for slow interruptions. For reasons stated above no shock is felt on making—but only on breaking, when a single contraction occurs. Now let the hammer vibrate rapidly (about 50 times a second): the muscles remain contracted during the whole time of the flow.

The "induced current" is made up of a series of very short currents, each of which acts as a galvanic make excitation only:[*] the effect of a galvanic current very rapidly interrupted, as for instance, by means of a toothed wheel, is also to produce persistent muscular contraction. In either case the "tetanus" is due to the fact that each successive stimulation falls upon the nerve before the muscle has reached its maximum contraction and begun to relax; so that a fresh stimulus is conveyed to the muscle in time to prevent its relaxation from taking place.

Though the faradic current has no marked electrolytic properties, its two poles are to be distinguished in physiological experiments.

EXPERIMENT XIV.—Place a commutator in the circuit and examine the effect of reversing the induction current:[†] the kathode is the more powerful of the two; and as in the case of the galvanic current does not excite exactly the same muscles, or parts of muscles,

---

[*] Faradic currents being of exceedingly short duration, and when of "medical" strength having no perceptible electrolytical action cannot give rise to anelectrotonus in their anodic zone of action. For we know that anelectrotonus requires a certain time to be set up, and depends in all probability on a certain chemical alteration of the nerve. It must be assumed, therefore, that they never produce any opening contractions.

[†] We designate here by "induction current" the succession of break currents only since the make currents, have just been shown to have no physiological effects. The reader must not get confused by the expressions "make" and "break" applied to induction currents. These expressions indicate their mode of origin; viz., makes and breaks of the induc*ing* current in the primary coil. Every make and every break induc*ed* current has a beginning and an end, a make and a break, and as just said excites nerves at *its* make only.

as the anode, though the electrode remains immovable on the same point of a nerve. Both these facts are illustrated by the rougher experiment of holding the kathode and anode in the right and left hand respectively; the effect on the flexors and extensors of the hand varies on the two sides, because the virtual kathodic zones are not exactly the same in the two arms.

EXPERIMENT XV.—Repeat the experiment on the voltaic alternatives (Exp. X) with a rapidly interrupted faradic current, which you reverse after it has flowed for some time in the same direction. You will find that the effect of such reversals is greater than that of simply breaking or making again that current.

Faradic excitations increase to a certain extent the excitability of the nerve. This is shown by myographic tracings of contractions obtained by series of single faradic shocks, when a more or less rapid increase is observed at the beginning of each series; a fact which proves the after effects of faradic currents in the kathodic zone. The experiment just made, on the other hand, seems to show the existence of such after effects in the anodic zone—though not so marked as those produced by the passage of a galvanic current, nor possibly of the same nature.

EXPERIMENT XVI.—Has shown that there is the same difference between the physiological effect of the faradic as between that of the galvanic poles. Estimate this difference by noting the extent of the contraction obtained with A and K. Now include in the circuit a resistance of 2 or 3 thousand ohms (a long piece of wet thread between the apparatus and the electrode does very well (cf. page 11). You will find that the effect of the positive pole is considerably diminished relatively to that of the negative pole. The difference between the two is as great as, or greater than, that between the galvanic poles. The explanation of this phenomenon is by no means clear.*

C. ELECTROTONUS.

Several attempts have been made to show by direct experiment on nerves within the intact living body, that the alterations of excitability during the passage of a galvanic current exist in them which are found in the exsected sciatic of the frog, and have hitherto been assumed by us to exist in the imbedded nerve, in order to explain its several reactions. The alterations consist as we saw of a state of exalted excitability at the kathode (katelectrotonus) and of a state of depressed excitability at the anode (anelectrotonus). The method of demonstrating them in the ex-

* It is possible that the resistance by delaying the flow of the current increases its diffusion. If it be true that the diffusion of faradic is less rapid than that of galvanic currents, resistance would equalize them in this respect. The peripolar faradic zone, (where the anodic make stimulus falls) would then occupy a larger area than before (i.e., the electrical density in it would be less dense), and the effect of that stimulus, ACC, would be diminished in comparison with KCC which starts from the polar zone in which the density does not vary in the same degree.

sected nerve is very simple. A stimulus of uniform strength (electrical, mechanical, or chemical) is applied to the nerve, in the neighbourhood of the points of application of the electrodes, before and during polarisation (*i.e.*, the passage of the galvanic current) and the effect estimated by the amplitude of the muscular contractions produced under the varying conditions of experimentation. It is thus proved that the nerve is more excitable in the kathodic zone, less excitable in the anodic zone, than in the normal state.

Experiments on the human subjects performed on the same plan could not yield any clear and consistent results, the physical conditions being entirely different, (see Diagram 78). We have shown the necessity of modifying the methods accordingly. Moreover means had to be provided for the elimination of numerous sources of error, and for obtaining an objective record of the results obtained. These matters have recently been experimentally investigated by Waller and myself.[*]

The leading facts of electrotonus in the human nerve may be demonstrated by the following experiments :—

EXPERIMENT XVII.—Connect the positive pole of a galvanic battery[†] with the negative pole of a secondary induction coil and attach the electrodes to the negative pole of the galvanic battery and the positive pole of the coil. Arrange the electrodes as in experiment I, the combined kathode resting upon the nerve. Find at what distance of the secondary from the primary coil the induced current begins to produce contractions. Next throw in a current of IC cells. The nerve now reacts at a much lower power of the faradic current. The excitation falls upon the polar zone which is kathodic: hence we conclude that the nerve is there in a state of increased excitability, or of katelectrotonus. Repeat the experiment successively several times so as to be sure that the diminished cutaneous resistance does not influence the result.

EXPERIMENT XVIII.—Repeat the experiment with the combined anode on the nerve. The same effect is observed, but not so marked, because here the kathodic polarisation and excitation fall upon the peripolar zone where the electrical density is less.

EXPERIMENT XIX.—Connect the negative pole of the faradic with the negative of the galvanic apparatus. Attach the small electrode to the positive pole of the coil, the large electrode to the positive of the battery, and fix the electrodes as in experiment I. The small electrode is thus faradic anode and galvanic kathode. Choose a strength of faradic current which gives slight contractions. Now add the galvanic current cell by cell. The effect of the faradic current is at first diminished sometimes to extinction. But as the galvanic current grows the contractions reappear, increase, growing even to above the normal.—In this experiment the faradic kathodic excitation falls upon the peripolar zone which is under the galvanic anodic influence. The diminished effect shows the

---

[*] *Philosophical Transactions of the Royal Society* 1882, p. 961. ff., plates 64 and 65. Abstract in *British Medical Journal*, Vol. I, 1882. The following experiments give the outline of the methods followed by us in this research.

[†] Of low internal resistance, *e.g.*, a freshly charged Leclanché.

diminished excitability (anelectrotonus) of the nerve in that region. The reappearance and subsequent increase of the effect shows that beyond a certain strength of polarising current there is as it were an invasion of the region of anelectrotonic alteration by the katelectrotonic influence. The excitability of the nerve is at first depressed then grows again under the increasing galvanic influence. In the frog's nerve it has been shown that a similar invasion occurs but in the opposite sense: the kalelectrotonus gives way, and the anelectrotonus advances with increasing current strength. We cannot explain the reason of the difference between the exsected and the embedded nerve in this respect.*

We conclude that electrotonic changes do occur in the human nerve as they do in the frog's nerve. This conclusion is borne out by other considerations as well as by experiments made with galvanic and mechanical stimuli, but far too complicated to be mentioned here.

We have seen (Exp. VIII) that the AOC grows (within certain limits) with the length of the anodic duration which has preceded it. We now see the reason of this fact: the more complete the depression of the excitability, the more energetically the release from it acts as an excitant. The gradual disparition of AOC by repetition (Exp. IX) may possibly be ascribed to the katelectrotonic invasion, possibly also to the agency of the "after effects" we have had the occasion of mentioning at page 108, which we shall now consider. By *after-effects* is meant the alterations in the excitability of a nerve left by a current after it has ceased to flow through it. On the frog's nerve they consist of, 1st a positive modification, that is an increased excitability of the nerve in the previously anelectrotonic region, 2nd a positive modification also, but preceded by an exceedingly short negative modification in the previously katelectrotonic region.

They are demonstrable in man, but require a great nicety of manipulation, and the use of the graphic method, so that the experiments showing them cannot be described here. Their nature however will easily be understood from what we have said previously of voltaic alternatives. There is one difference however worth recording between the phenomena displayed by the frog's and by the human nerve: in the latter the negative modification after the cessation of katelectrotonus is well marked and may last several seconds after sufficient polarisation. Under the same circumstances the positive modification is of long duration. We have traced it for nearly two hours in one experiment.

We need not say much of the attempts made to investigate the electrotonic changes in the human nerve by testing, not through the polarising electrode itself as we have done in the experiments just described with a view to secure the coincidence of the zones of

---

* This katelectrotonic invasion may perhaps account for the comparative ease with which KCT is obtained in man. In the frog's nerve, on the other hand, the AOT is well marked after strong polarisation, and anelectrotonic invasion. The apparent discrepancy between the results on the human and the batrachian nerve may be possibly considered as the manifestation of a deeper parallelism.

polarisation and excitation, but through a second electrode in the neighbourhood of the former as is usually done in the case of the exsected nerve. The results are, as we might expect, subject to variations according to the exact spot where the excitation falls. Thus the diagram (fig. 78) makes it evident that according as the stimulus falls in the polar or in the peripolar zone we shall have evidence of either diminished or increased excitability, as the case may be.*

MUSCLE.

The electro-physiology of muscles is to be studied by performing on them the same experiments as have just been described with reference to motor nerves. For this purpose the electrode to be applied on the muscle may advantageously be of medium size when one of the larger muscles is to be excited (*e.g.*, extensor femoris gastrocnemius, etc.). We need not say more with reference to the results to be obtained than that on healthy muscles they are governed by the same conditions and follow the same laws as those derived from the investigation of motor nerves. Muscles however being organs of a more complex structure than nerve fibres, are less suitable for strictly physiological experimentation whilst their behaviour in pathological conditions makes their examination the chief task of electro-diagnostic inquiry. We shall therefore confine ourselves here to a demonstration of the "motor points," and reserve a few remarks on the reactions of healthy muscles for our chapter on electro-diagnosis.

As in the case of nerve excitation the observer must carefully observe and learn to recognise the mode of contraction of healthy muscle when directly excited (see page 99). We shall have the opportunity of returning to this point when we come to describe the alterations in the mode of response of diseased muscular tissue.

EXPERIMENT XX.—Take a faradic apparatus; connect to one of its poles a large electrode which is kept fixed to the trunk of the subject; to the other pole a small carbon or sponge electrode, the handle of which you hold in the hand. Let the current be of such strength as to provoke moderate localised contractions of the superficial muscles. Move the electrode over the surface of a number of muscles such as the sternomastoid, abductor indicis, deltoid, triceps extensor, extensors of the hand and fingers, peronei etc. You will soon perceive that now and then the contractions of the muscle examined appear to be much intensified. Mark with ink the spots where the electrode happens to be when the increased effect occurs. Compare the subject thus marked with the figures given of the "Motor points." If you have made no mistake you will find that your markings correspond with the dots on the figure.—Note parti-

---

* Our own results agree with those of Erb on the subject, and confirm the theory of the two zones; but we found that, as was to be anticipated, the effects varied also with the pole used for testing. But the discussion of these matters would lead us beyond the scope of an elementary book like the present.

cularly the great difference produced in the amount of contraction by a slight displacement of the electrode from the "motor point." Test and mark in the same way the various superfical nerves of the face, neck and limbs, noticing the groups of muscles brought into play.

The fact that muscles react more readily when excited at certain points (of "election" as Duchenne called them) had been known for some time before Ziemssen investigated the question thoroughly and proved that such points correspond to the spot where the motor nerve enters the muscle. When the electrode rests over such a spot it is obvious that the whole mass of the muscle is excited at the same time; whilst in any other position the electrode acts only upon the individual muscular bundles subjacent to it (owing to the rapid diffusion of the current through the muscle). You will notice that a number of round muscles have more than one "motor point," the deltoid for instance, the gastrocnemius, etc. Flat muscles, obliqui, serrati, etc, have a larger number still so that it is not possible to bring them to a contraction "en masse" by intramuscular stimulation. In such cases you must use large flat electrodes and stronger currents; or if possible get at the nerve trunk which supplies them, as for instance the posterior thoracic nerve, above the clavicle or at the axilla, in the case of the serratus magnus.

The methods of exciting muscles at their motor points, and through the nerve trunk, have been called direct and indirect, respectively. The direct method has been brought to a high state of perfection by Duchenne under the name of "Electrisation localisée." So highly did its originator prize this fruit of indefatigable labours that he named after it the work in which are recorded his still more lasting titles to fame, namely his clinical and pathological descriptions of locomotor ataxia, infantile paralysis, and progressive muscular atrophy. The value of "localised electrisation" itself, however, lies not in its therapeutical, but in its physiological applications. It was by its means that Duchenne was able to raise another monument worthy of his genius, the "Physiologie des mouvements" and "de l'expression." The isolated, artificial, contraction of many muscles on the living subject could alone solve the problems of their functions which had baffled the ingenuity of anatomists during centuries.

From what experimental researches have taught us concerning the reactions of unstriated muscular fibres to electrical currents and from a few manometric experiments as well as numerous clinical observations on the human subject we may assume that with appropriate methods, most of the muscular abdominal organs may be influenced by percutaneous electrisation. The unstriated muscles react to polar influences as do striated muscles, but their contractions are slower and more durable. Ziemssen has quite recently shown upon the exposed human heart that this organ does not react to induced currents applied to its substance. To galvanism however the heart responds like all other muscles. A quickly interrupted galvanic current brings the cardiac rhythm up to its own frequency (160-180 a minute); a slowly interrupted current retards the heart's action, but not very readily (50 a minute). These phenomena may be observed on percutaneous galvanisation of the normal heart.

SENSORY NERVES.

Throughout the experiments previously made on motor nerves (*i.e.*, mixed nerves) the observer will have noticed that sensations accompanied the application of electric currents to the body. These sensations, on analysis, are referable to two categories: sensations, perceived under the electrode itself; sensations referred to the distribution of the nerve. Each of these categories again may be subdivided: the first contains the sensations due to the excitation of the cutaneous nerve endings, and those to the electrolysis, or chemical action set up by the current, in the skin; the second embraces the sensations of the muscular contractions produced by the current, and those due to the excitation of the sensory nerve fibres contained in the excited nerve trunk. It is with the last only that we need concern ourselves when trying to determine the influence of currents on sensory fibres.

In order to eliminate the confusing effects of concomitant muscular contractions we choose a nerve, such as the supraorbital, which contains no motor elements. The different localisation of the sensations produced under the electrode itself, and of those referred to the parts of distribution, makes it easy to discriminate between the two kinds.

EXPERIMENT XXI.—Fixing a small electrode over a sensory nerve (supraorbital, digital, etc.) which can be readily excited without producing any muscular contractions, whilst the other electrode of large size remains securely fixed upon a distant part of the trunk, make and break the galvanic current, proceeding as in Experiment V. and noting at what strength of current sensory irradiations occur when the negative and when the positive poles are on the nerve.

You will find that sensory nerves react in the same way as motor nerves; that is most readily to kathodic make (K C), then to anodic make and break (A C, A O), finally to kathodic break (K O). In other words the so-called "law of contractions" of motor nerves is applicable to the sensory nerves the formula for which may be written 1. KCS; 2. ACS, AOS; 3. KOS; where S stands for "sensation." There petition of Experiments XVII etc., on sensory nerves will show that the electrotonic alterations of their excitability follow precisely the same course as has been observed in motor nerves.— It will be noticed that the excitation of sensory nerves during the whole time of flow (apart from the local burning cutaneous sensation), and which corresponds to the duration tetanus (D T) of motor nerves, occurs with very low current strengths. This fact makes the O S somewhat difficult to recognise if the current is strong.

The effects of the attention are a fruitful source of fallacy in all experiments on sensation, in which the record is purely subjective. The minimal excitation perceived varies according to the method used to determine it. It occurs with a lower current strength when sought with a gradually diminishing current, than when sought with one made to increase from a point below the minimum.

EXPERIMENT XXII.—Take a small electrode, keep it applied with perfect steadiness upon the closed eyelid whilst, the other electrode of large size rests upon a distant part of the body. Begin with a current of two cells, which you gradually increase by 1 or 2 cells at a time; at each increment make and break the current with a uniform rhythm. Do this when the negative and when the positive pole is on the eye, and note the general characteristics of the optic reactions to KC, KO, AC, AO—(intensity, colour, persistance etc). The optic apparatus reacts to all excitations by a manifestation of its specific energy (*i.e.*, by a conversion of the stimulus into an impression of light), according to the physiological law common to all nervous structures. The sensation experienced with moderate galvanic shocks is that of a transitory or persistent illumination of the field of vision, comparable to the phosphenes obtained by mechanical excitation of the eye. On repeating several times the experiment, (which is perfectly safe, even with much stronger currents than one would imagine from the accounts met with in certain electrotherapeutical writings), and applying oneself to the analysis of the phenomena, one discovers that the subjective flashes of light are not all of the same colour. Those occurring at KC and AO differ from those occurring at AC and KO, each pair of excitations giving rise to the same colour or combination of colours. On experimenting on a large number of individuals one finds this fact to be constant, though the colours perceived by each individual are not the same for all. Thus for some the electrophosphenes are reddish and blueish, for others yellowish and blueish, and so on. Instead of a simple coloured field, again, there may appear a disc with a blueish or yellowish centre, surrounded by a yellowish or blueish periphery respectively; but here also the colours differ with the polar excitations which give rise to them (KC and AO differing from AC and KO phosphenes, but in each pair the phosphene being alike).

Now in Exp. VII, we have described observations on the motor nerves which yield results too obviously parallel to those just described, not to suggest an analogy between the conditions giving rise to the phenomena in each case. We have seen there that the muscles excited by KC and by AC differed; and we have explained the difference on the theories that the excitation in each case fell in the polar and peripolar zone respectively; and that the effect of AO, under favourable conditions, was the same as that of KC because the excitation in each case fell in the same zone (polar). It is obvious that the same explanation might account for the phenomena as observed in the excitation of the retina and optic nerve: different parts of those structures being excited by KC and AO on the one hand and by AC and KO in the other, the results in the two cases are not the same. Evidence confimatory of this view is obtained by placing the electrode on the temple instead of on the eyelid. The phosphenes are then seen to travel across the field of vision, either from within outwards or vice-versa; and if persistent, appear to settle in the inner or outer half of the retina—and in both these particulars the effect of KC or AO differs from that of AC or KO.

EXPERIMENT XXIII.—Place a medium sized electrode just in front of the ear so as to cover the tragus (but *not* to close the meatus); proceed in other respects, as in the examination of the motor nerves (experiment I) or of the eye (experiment XXII). You will find that a healthy acoustic nerve reacts only to KC and to AO (*i.e.*, to excitations falling in the polar zone) the reaction consists in a sound variable as to its pitch and duration. It may be of a ringing, hissing, humming, buzzing, etc., character.

Experiments on the electrical reactions of the acoustic nerve offer certain difficulties. Many individuals require such strong currents to elicit a response that the giddiness and other unpleasant symptoms accompanying such excitations preclude us from obtaining on them satisfactory results. In others, however, the acoustic reactions are elicited with weak currents, and may therefore be studied in detail.

The acoustic reactions are of twofold interest. First historically since it was through their study that Brenner was led to insist upon the "polar method" of applying the current to the human body.* Second, physically, as they seem to be confirmatory of the polar and peripolar zone theory. By reacting to KC and AO only the acoustic apparatus shows itself excitable only in the polar zone, the reason of this exceptional circumstance being probably to be found in the peculiar disposition of the nervous elements with reference to the badly conducting material (bone) with which they are surrounded. The latter prevents a rapid current diffusion, and hence the formation of a peripolar zone of sufficient density for an electrical excitation of the nerve.

A very characteristic phenomenon attending the circulation of even very weak currents through the gustatory organs is the peculiar "galvanic" taste experienced, not easily described, but not readily forgotten.

EXPERIMENT XXIV.—Apply the poles (middle-size electrodes) of a current of 6-10 cells to the two cheeks. You will notice that the sensation of taste is much stronger on one side than on the other, and on investigating the polarity of the electrode resting on that side will find that it is anodic. This taste is of a metallic, sub-acid, nature. On the kathodic side, besides being weaker, the sensation is somewhat different in quality. Perhaps the explanation of this fact is that the kathodic excites the tactile rather than the gustatory nerve-elements, the latter being perhaps more readily influenced by the chemical changes set up at the anode.

EXPERIMENT XXV.—Place an electrode on the nape of the neck, or on some other part of the body, and touch the tip of the tongue with the other, making it alternately kathode and anode. You will find that the sensation of taste is more vivid when the kathode

---

* The polar reactions of motor nerves had already been described by Baierlacher and Chauveau as I have shown elsewhere. (*Brain*, 1880).

touches the tongue, but it then occurs not under the electrode but at the back part of the tongue and fauces. This fact is another confirmation of the theory of the zones: the back of the tongue is the part best supplied with gustatory organs, and is occupied by the anodic zone when the kathode rests on the tip of the tongue. Hence the sensation will be more intense in the latter case than when the tip is under anodic influence.

EXPERIMENT XXVI.—Send a current gradually increased to 10 cells or more, through the head with the electrodes first on the occiput and forehead, then on the two temples, finally on the two mastoid processes. You will find that the effect in the first case is confined to an undefinable, sickly, sensation of cerebral disturbance whilst in the latter two cases giddiness, culminating in loss of equilibrium, is experienced. The more the direction of the current through the brain departs from the longitudinal to assume the transverse direction, the more readily giddiness supervenes. If the current be very strong the eyes are turned towards the negative pole, and perform slight movements like those of nystagmus.

EXPERIMENT XXVII.—Make and break a current from 6 to 12 cells (or more if necessary) whilst the electrodes rest on the mastoid processes. A tendency to fall to one side is objectively observable, whilst the subjective impression of such a fall is vividly experienced. The fall occurs from the kathode on making the current, from the anode on breaking it, that is, to that side of the body which is under the control of the excited cerebral hemisphere. These phenomena are probably due to an interference with the centripetal influences necessary for the perservation of the equilibrium of the body (Erb). The movements of the trunk are of the same character as the "forced movements" observed in animals under certain experimental conditions, being due to actual muscular contractions on one side of the body. Hitherto it has not been possible to bring into action the cortical centres by means of percranial excitations. Certain observers have described vasomotor phenomena in the encephalon of animals subjected to percranial galvanisation;[*] but it is open to great doubt whether currents of the necessary strength to produce such changes can even be applied to a human brain.

Physiological experiments on vasomotor nerves, including the cervical sympathetic, cannot be performed on the human body. At any rate the numerous attempts hitherto made have conspicuously failed to give such results worthy of any serious consideration. Here and there observations, as exceptional as they are conflicting, granting they are reliable, have been recorded, and the most preposterous claims set forth on their behalf, as if they furnished grounds sufficient for drawing physiological, or therapeutical, conclusions. Electrical experiments upon the spinal cord have hitherto

[*] Löwenfeld says that when the current is transverse there is dilatation of arterioles on the positive, contraction on the negative side, when longitudinal there is contraction if the current be descending (positive pole on forehead, negative on neck), dilatation if ascending.

failed to yield any information available for guidance in practice. The same may be said concerning secretory and trophic nerves—or rather concerning the influence of the nervous system upon secretion and nutrition.

APPENDIX.—POLARISATION OF THE HUMAN BODY.

Allusion has been made at page 45 to the change in the current-strength—a marked augmentation—observed when a current is reversed, after it has passed for a short time through the body. This increase obviously plays an important part in the greater effect of a voltaic alternative than that of a simple closure (as mentioned at page 108) and should be eliminated from our experiments (by the method described, at page 39), if we wish to observe the effect of the "positive modification" (p. 112) on nerve excitability.

I had reserved the discussion of the phenomenon for this chapter in the hope that by the time it was going through the press, I should have been able to mention the results obtained by Waller and myself. But our experiments not having yet been concluded, I am compelled to confine myself to a statement of the leading facts observed, (cf. our paper, previously cited, in *Phil. Trans. Royal Society*, 1882), without attempting to explain their nature.

When a current is reversed, after flowing for a certain time through the body, the galvanometer shows a considerable increase in its strength. This increase may be due either to a diminution of resistance or to an increase of electromotive force in the external circuit, *i.e.* in the body and electrodes. That it does not depend upon the latter circumstance, is shown by the fact that the electrodes and body connected to a galvanometer do not give any such current, as would be the case if their polarisation had developed an electromotive force sufficient to account for the increased current on reversal. That it does not depend upon a simple diminution of resistance due to the permeation of the skin, as assumed by the few electrotherapeutical writers who mention the phenomenon, is shown in that the increased deflection of the needle is not permanent, but soon falls off to return to the former maximum or thereabouts. The rapidity with which the needle so returns, is greater the shorter the time of previous polarisation has been.

Chapter III.

ELECTRODIAGNOSIS.

ELECTRODIAGNOSIS AND PROGNOSIS OF MOTOR NERVE AND MUSCLE LESIONS.

We have studied in the preceding chapter the reactions to electricity of the healthy human nerve and muscle. We have shown how the apparently anomalous phenomena harmonize with the data of experimental physiology when we interpret them in the light of the various physical conditions under which these various results are obtained. There is a "law of contractions" for the human muscle explored by the method of unipolar stimulation; that is to say the readiness or energy with which a muscle answers to galvanic stimuli differs according to the pole, kathode or anode, applied to it or to its motor nerve, and to the change, closure or opening, brought about in the circuit. We have also noticed that a constant current of a certain strength gave rise to persistent contractions during its flow, (page 99).

Adopting the letters K and A to designate the polarity of the electrode over the nerve or muscle to be tested (Kathodic or Anodic); C and O to designate the nature of the electrical oscillation (Closure or Opening) at which the contraction (C) occurs, we have expressed the order in which the reactions are obtained, as (1) KCC, (2) AOC and ACC, (3) KOC. In this series is to be intercalated the supervention of tetanic contractions (T) observed during the flow (D, or Duration) of a sufficiently strong current. KDT and ADT occur usually, the one before, the other after the apparition of KOC. The tabular arrangement of the contractions obtainable on a healthy subject, and arranged in their chronological order of apparition with increasing current strengths, is known under the name of "the normal polar formula," or "laws" of contractions, (p. 100).

In disease the excitable tissues are found frequently to have lost their property of reacting to electrical stimuli in a normal manner. There may be exaggeration or diminution of their response; or there may be a disturbance in the order of their contractions as compared to the normal formula.

Now experiment has proved that some of those changes have a well-defined connection with certain alterations in the nutrition of the neuro-muscular apparatus—in fact that the latter stood to the former in the relation of cause to effect. In other words, that we might from the occurrence of certain irregularities in the amplitude, mode, and order of their polar responses, infer a degenerative process in the nerves and muscles. The most important features are those displayed by degenerating muscle; for besides yielding an abnormal formula, such a muscle differs from a healthy one, inasmuch as whilst at first it displays an exalted excitability to slow galvanic makes and breaks, it loses its excitability to short stimuli, such as the faradic or the rapidly interrupted galvanic current; in addition to which changes we find that the mode of normal contraction is altered in such a muscle. This alteration denotes a degradation of the striated fibre to a lower type of contractile tissue, such as that found in muscles of organic life and in lower organisms, like molluscs. The electrical stimulus, instead of a well-defined, short-lived jerk, calls forth a lazy, long-drawn, tonic wave, the summit of which is low relatively to the time taken in its ascent and descent.

The reactions of nerves which are the seat of abnormal nutritive processes also present differences from the response of healthy nerve; but I need not anticipate here upon the description to be presently given of these phenomena.

It is upon these elementary facts that the art of electrodiagnosis is founded, which in many cases of neuro-muscular disturbance yields to the physician information as important as does chemical analysis in cases of renal disorders, and physical examination in cases of thoracic disease.

The popular theory is that the carrying out an electrical investigation requires no special training, and is within the reach of any one possessing a battery and a superficial knowledge of the medical applications of the current to the human body. What should we think of the pretension of a second year's student to being able to make a satisfactory stethoscopical diagnosis? And yet is not the former notion as absurd as the latter? In electrodiagnosis are involved problems, the solution of which requires a full mastery of the principles of electrophysiology, as well as a competent knowledge of the neuropathological facts which underlie its phenomena. The fallacies besetting it are far more numerous and subtle than those attending the other methods of physical examination of the living organism. Electrodiagnosis is a branch of experimental science, and science is no respecter of persons. The ablest physician, the most consummate neurologist, if he has not submitted himself to the conditions required, and passed through the ordeal of discipleship, will fail to obtain reliable results. The statement of such results, even when expressed on oath in a court of law, is devoid of the least weight, and as experience has shown, more likely to be wrong than otherwise. It is less than an opinion—it is a mere guess. To be the expression of a scientific fact, it should have been reached by scientific methods, in which every source of fallacy had been carefully eliminated.

It is true that often enough the results of an electrical investigation are grossly palpable and not likely to be overlooked by the most casual observer. But even in such cases fallacies are possible, as proved by the astounding mistakes that have been made, even by men of high repute; and after all such instances are not those where the diagnostic value of electricity is greatest. Just as the facts elicited by the physical examination of a heart or lung in the last stage of disease, do not reveal much beyond what the general aspect of the patient has led us to expect; so likewise but little additional light is shed upon the practical issues of the case by the reactions in a typical instance of infantile or lead paralysis, or of advanced progressive muscular atrophy. Where electrodiagnosis is of paramount importance is at the beginning, during the insidious stage, of numerous neuropathies. Here, as in the earliest stage of phthisis for instance, the discrimination of subtle differences between the normal and abnormal signs, requires that special training which cannot be expected from every member of the medical profession. At the same time the latter ought to be fully aware both of the importance and difficulty of the problems. Attempting to cut the Gordian knot by a clumsy blow of a blunt sword; and sneering at the alleged pretensions of the "mystery-men" to keep a monopoly of their art, are both to be deprecated as irrational and undignified.

The results of an electrical examination of motor nerves and muscles may establish one of three things:—

(1.) The nerves and muscles examined react *normally*.

(2.) Their reactions present *quantitative* abnormalities—that is to say, there is either an excess or a deficiency in the amplitude of the contraction following an excitation of given intensity. (*a.*) The "minimal" contraction occurs with a strength of current less or greater than that required to bring it about in the healthy organ. (*b.*) The "maximal" contraction obtainable in health cannot be elicited even with the strongest currents. When alone present the latter condition depends more upon deficiency of contractile tissue than upon deficient excitability.

(3.) They present *qualitative* abnormalities—that is to say, (*a*) the order of apparition, or relative amplitude, of the muscular contractions to the kathodic and anodic make and break of the galvanic current is altered ("serial" alteration); (*b*) the mode or course of the contraction, *i.e.* the myographic curve, is altered ("modal" alteration). When qualitative alterations are present, quantitative changes are invariably present also. This combination of quantitative, serial, and modal alterations in the muscular response is known as the "Reaction of Degeneration" (RD).

The only direct inference to be derived from the reactions of nerves and muscles to electricity relates to the state of their nutrition. If these reactions are abnormal, their nutrition also is abnormal. When speaking of the reaction of degeneration we shall have the opportunity of dwelling more fully upon this important topic. In the meanwhile we may content ourselves with bearing in mind, first, that the normal nutrition of nerves and muscles indicates a healthy condition of the nerve-centres in the anterior spinal cornua, and of the nerve-fibres themselves, from the ganglionic cells composing those centres, down to the explored region; secondly, that such a normal nutrition is consistent with very considerable disturbance in the functions, whether in the direction of excessive or of diminished motility, of the organs examined. Such a disturbance may depend upon various lesions of the higher nerve-centres and paths.

I. *Normal reactions.*—The diseases of the neuro-muscular apparatus in which the reactions are found to be normal are chiefly:—

(1.) Cerebral disease or injury affecting the cortex and its fibres; the corpus striatum, pons, or peduncles. In other words all paralyses and spasms of cerebral origin, whether due to organic hæmorrhage, embolism, &c.), or functional disturbance (hysteria, epilepsy, chorea,° &c.), are usually unaccompanied by any alteration of the electrical reactions as long as no secondary changes are set up in the cord.

(2.) Spinal diseases affecting the white matter only, frequently do not influence the electrical excitability, at least during their first stages: such are locomotor ataxy, lateral sclerosis, disseminated sclerosis.

(3.) Spinal lesions affecting a circumscribed portion of the cord in its whole thickness give rise to no electro-diagnostic phenomena in the parts below the part affected (transverse myelitis, compression, crushing, &c.). The nerves and muscles under the direct influence of the diseased portion will, however, present the reactions of degeneration.

(4.) Certain morbid states of the peripheral nerves, such as those obtained in the mildest cases of facial or musculo-spiral paralysis from cold or pressure, may be unaccompanied with any departure from the normal excitability.

We have already said the only direct conclusion to be drawn from the fact that the electro-excitability† is normal, is that the nerves and muscles which are the seat of the alleged diminution or excess of motility (paralysis or spasm, hypo- or hyper-kinesis), have not suffered in their nutrition. This conclusion is consistent with the co-existence of any one of the morbid conditions enumerated above, but does not exclude the possibility of shamming on the part of the individual examined. In order to determine the latter point we shall have to consider whether the history and other symptoms of the case tally with one of the diseases shown to be possible by the electrical investigation. If we are satisfied that the case is genuine, we are enabled by the discovery that the reactions are normal to exclude the participation of the spinal grey matter or of the nerve-trunks in the production of the symptoms. In some cases no doubt, the clinical aspects of a case dispense

---

* I am aware that some authors have ascribed increased excitability to the nerves of choreic and other patients. But as their observations were apparently conducted without reference to the resistance of the parts tested, and as in those cases where the test was applied under galvanometric control, the apparent increase was shown to depend upon physical causes, we may relegate their statements with a vast mass of similar allegations to the class of errors of observation.

† The term *electro*-excitability should be carefully restricted to its generic sense, including within it both *galvano*- and *farado*- excitability, or contractility. I do not in the least intend to convey, by these several expressions, that there is a special property in nerve and muscle by which they react to electrical stimuli, but use the terms in order to avoid the fastidious repetition of a longer, though more correct, periphrase.

altogether with the necessity of consulting the electrical reactions, or restrain its utility to the determination of any secondary changes which may have supervened in the cord or peripheral nerves. But in not a few cases the presence or absence of electro-diagnostic signs is of the highest significance, and give the only clue to a true interpretation of the symptoms. Thus, to give but one example, I shall mention the discrimination at an early stage of cerebral paralysis (cortical, hysterical) from true spinal (poliomyelitic) disease. On the other hand the reader will see how unfounded the notion, which is still current among many physicians and surgeons, that an electrical investigation is a ready method of distinguishing between organic and functional disease, *e.g.*, between hysterical and leucomyelitic paraplegia.

II. *Quantitative changes.*—If the excitability is (*a*) *augmented*, the nerves and muscles reacting to a lower minimal stimulus than normally, or the contractions being excessive to a given current strength, we may indeed conclude to some actual change in the molecular state of the nerve, but the information gained is not of any great practical importance, as the nature of that change is entirely hidden from us. It is supposed to depend in many cases upon an irritation or exalted reflex activity of the anterior cornua of the cord. Hyperexcitability is found in the early stages of cerebral hæmorrhage, of locomotor ataxy, of facial paralysis, &c., but the phenomenon is not constant, especially in the hands of observers who eliminate the effects of altered body resistances by means of a galvanometric control. It is, however, extremely well exemplified in tetanilla, where the peripheral nerves have also their mechanical excitability exalted to a very remarkable degree.

Increased excitability of muscles to the galvanic excitations is a characteristic feature of the early stages of degenerative reactions, as we shall presently see.

(*b*). *Diminution* of the excitability (besides being a very constant accompaniment of the RD in several of its phases) occurs under numerous conditions, and may culminate in total abolition of response. Old-standing cerebral paralyses, involving secondary descending changes, and still more, diseases of the spinal white matter in their later stages, are characterised by a deficiency in the neuro-muscular excitability. The same may be said of certain cases of very slowly progressing muscular atrophy, in pseudo-hypertrophy of the muscles, and perhaps in the milder forms of idiopathic neuritis. Simple muscular atrophy from disuse, wasting diseases, or affections of the joints, is accompanied with deficient response to maximal rather than to minimal excitations.

From this enumeration it is obvious that the chief conclusion to be drawn from the occurrence of diminished electrical reaction is more than a negative one. It excludes on the one hand poliomyelitis, on the other hysteria and all purely cerebral disturbances, as well as shamming. It invariably points to certain alterations in the nerve-fibres. In numerous cases it enables us therefore to prove the organic nature of a disease, and assists in localising and comparing the depth of the lesion.

In such cases, where the other features of the disease are not well marked, and where at the same time the diminution of excitability is slight, the utmost care in applying the principles of investigation laid down previously must be taken before we can place any reliance upon the results; for, having here no qualitative changes to assist us in forming an opinion, we depend entirely upon an accurate dosage of our exciting agent in the diseased as compared to the standard healthy nerve.

We do not know exactly the nature and amount of organic changes in the peripheral organs to which diminution of excitability is to be attributed. Nor can any reliable prognostic conclusions be always founded upon the degree of that diminution. It is important, in connection with this subject, to remember that a nerve which has regained its functional activity—after complete section, for instance—may remain for months or years without regaining its electro-excitability below the point of lesion, though perfectly able to transmit volitional or artificial impulses from above. We know that such a nerve has undergone a process of degeneration and of regeneration, but it is not yet perfectly clear why the latter stops short of restoring to the nerve its electro-excitability. It is scarcely permissible to conjecture as yet that the changes in peripheral nerves following old central (cerebral or leucomyelitic) lesions are of the same nature as this post-regenerative alteration. Still, such a possibility suggests itself almost involuntarily to the mind. The fact, however, remains incontrovertible, that diminution of excitability is due to some alteration of local or central origin in the normal constitution of the nerve.

III. The reaction of nerves and muscles present quantitative and qualitative (serial and modal) abnormalities: *Reaction of Degeneration* (RD). The phenomena attending excitation of the nerves and excitation of the muscles follow widely different courses, so that it is necessary to consider them separately. The appended diagrams will assist the reader in following the description of the somewhat complex phenomena of RD, (fig. 79, ff.)

(A). *Alterations in the reactions of nerves.* These consist mainly in a more or less rapid diminution of excitability (to both currents equally) according as the progress of the morbid changes is quick or slow. When the functions of the anterior horns of the cord are are suddenly suspended, as in infantile paralysis, or when a peripheral nerve is divided accidentally, or becomes the seat of a circumscribed acute process, as in ordinary facial paralysis, the nerve becomes unexcitable below the point of the lesion within two weeks. When the original disease is developed more gradually, as in chronic polio-myelitis or neuritis, the loss of nerve-excitability proceeds synchronously with it, and may continue to progress for many weeks or months.

The nerve reactions rarely show any qualitative changes; in certain cases, however, it is found that the muscular contraction obtained indirectly by galvanising or faradising the nerve are modally affected—that is, lose their character of an instantaneous twitch, and become sluggish as well as feebler. Serial modifications in nerve response have also been observed, but for the present must be looked upon as exceptional occurrences.

The period during which a nerve remains unexcitable varies considerably. In cases where, owing to the depth of the lesion, recovery does not occur, excitability never returns; but in certain cases, as we have already stated, it remains in abeyance permanently, or at least for a long time after the nerve has been restored to its physiological integrity as an organ of transmission. Usually, however, the return of electro-nervous excitability follows closely upon the first manifestations of that integrity—*i.e.*, upon the first transmissions of voluntary impulses.

(*B*). *Alterations in the reactions of muscles.* On exciting a muscle directly with (*a*) *faradism*, the phenomena are found to follow a course absolutely similar to the one they take in the nerve itself. This fact is fully concordant with the theory that by direct faradic stimulation of the healthy muscle it is really the intra-muscular nerve elements, and not the muscular fibres themselves, which are excited. Muscular substance requires stimuli of longer duration than the instantaneous fluxes of induction for its excitation. The disappearance of farado-muscular excitability is synchronous with and depends upon the degeneration of the intra-muscular nerve elements.

The effects of (*b*) *galvanic* excitation of muscle are highly characteristic. (1). *Quantitative* changes are present. Where the morbid process is acute there occurs first a slight diminution of response, soon followed by a well-marked rise. This rise is particularly well seen in severe facial paralysis of "rheumatic" origin, and in traumatic peripheral paralyses. In such a case the rise begins during the second week of the disease, and rapidly reaches a point where the weakest galvanic current (two or three cells, as against eight or nine require to produce contraction on the healty side) is enough to excite the muscles. This galvanic hyperexcitability persists for several weeks, gradually sinking back to normal, or below normal, according to the depth of the nutritive changes in the muscle. In chronic cases the diminution is the only quantitative alteration observed in the muscular response.

(2). *Qualitative* changes characterise the muscular contractions at the same time. These changes are both *modal*, the contraction being of a different type than in health, sluggish, with long ascent and descent, maximum shortening diminished, and *serial*.

The diagnostic value of (*a*) the *modal* changes is not often found adequately recognised (when it is not completely overlooked) in electrotherapeutical writings; and yet it is quite equal to that of the other modifications of muscular response to galvanism. In many cases a protracted contraction wave is the only ground upon which we can base our judgement as to whether RD is present or not.[o] This sluggish course of the muscular wave insensibly passes into the tonic contracton to be mentioned presently, the muscle remaining contracted during the whole time the current is flowing through it.

* The exaggerated mechanical excitability of muscular tissue presenting the phenomena of RD has been noted by Erb and Hitzig. This property probably depends upon the same histological changes which give rise to RD.

(*b*). *Serial* changes consist chiefly in the overtaking of KCC by the ACC. Instead of requiring, as in normal muscle, say, fifteen cells to obtain ACC where eight or ten give the first KCC, you require only twelve, ten, or fewer cells. A similar inversion is sometimes noticed between AOC and KOC, the latter being more easily obtained than the former, but the opening contractions usually become fainter, and disappear altogether within a short period after the disease has set in. KDT and ADT, on the other hand, become very marked, the latter especially, the muscle remaining in a state of persistent shortening during the whole duration of currents just strong enough to produce a closure contraction.

The length of time during which the qualitative alterations of muscular response may be observed varies with the general course of the disease. If recovery never takes place, the galvano-muscular excitability goes on sinking for a year or more; traces of ACC to strong currents may be elicited for a very long period in the last remainder of the dying muscular tissue. (Diagram 81).

Just as voluntary stimuli find their way through nerve-fibres still unexcitable to faradisation, so muscle may go on showing phenomena of RD when fit to obey the command of the will. RD may likewise be demonstrated in muscles which have never yet shown any signs of impaired functions, such as for instance in lead-poisoning, the extensors of the hand before any sign of wrist-drop is noticeable. Indeed, it cannot too explicitly be stated that paralysis and RD are symptoms between which there is no essential relation.

What does RD, then, signify? What are the pathological conditions to which it corresponds, and the alterations of nervous and muscular tissue to which it is due? It is of importance that we should have perfectly clear ideas on this point. The occurrence of RD depends upon a specific histological modification of the irritable tissues called "degenerative atrophy" (not idiopathic myositis, not simple atrophy or "wasting"), and this degeneration itself is due to an interference with, or a stoppage of, the peculiar influence of the grey matter (bulbar nuclei, anterior horns of the cord) upon the nerves and muscles, known as the "trophic" influence. Hence, from the occurrence of RD, whether complete or partial, we conclude to an alteration either of those centres themselves, or of the channels which convey their influence. These channels are the motor fibres, as proved by the results of experimental pathology.

According to the latest researches, section of a motor nerve is rapidly followed by coagulation of the medullary sheath, softening of the axis cylinder, and the breaking up of both, which eventually become reduced to a homogeneous mass of protoplasmic material filling the sheath of Schwann. The axis-cylinder does not survive, as was formerly believed, except when the nerve injury has been very slight. The neurilemma does not remain passive, but proliferates, and a cirrhotic condition of the nerve ensues. The intimate processes connected with the regeneration of nerve fibres have not yet been elucidated.

Whilst the sectioned nerve is going through these phases the muscles it innervates are the seat of corresponding histological changes. The fibrillæ lose their distinct striation, and apparently undergo an alteration in their chemical composition. There is proliferation of nuclei, and of the connective tissue, leading also to a cirrhotic condition of the muscle. The regenerative process begins when the nerve has been restored sufficiently to exert its influence upon the muscular fibre.

There is every reason to believe that the changes mentioned and just observed in warm-blooded animals, occur also in man under similar circumstances, and in idiopathic diseases of the trophic centres and of the nerve trunks. This probability is based upon the results of human morbid anatomy, and upon the fact that the course of the alterations in the electro-excitability both in man and in animals is essentially the same. The merit of having shown the dependence of the several phases of degenerative reaction upon the definite phases of histological changes, just described, belongs to Erb, who has embodied and summarised the facts in the annexed diagrams.

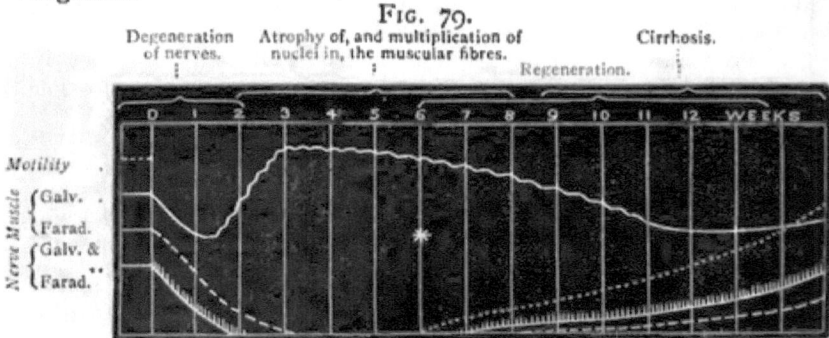

FIG. 79.

DIAGRAM A.   RECOVERY RAPID.

[N.B. The slope of the line indicating the fall of farado-muscular contractility should have been made abrupt so as to end near the bottom of line 2].

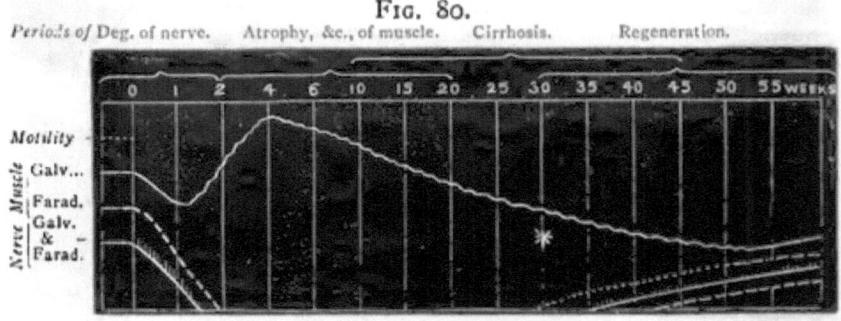

FIG. 80.

DIAGRAM B.   RECOVERY SLOW.

Diagram showing the synchronous evolution of phenomena, histological and electro-diagnostic, in a case of severe trophic lesion of the neuro-muscular apparatus, with slow recovery.

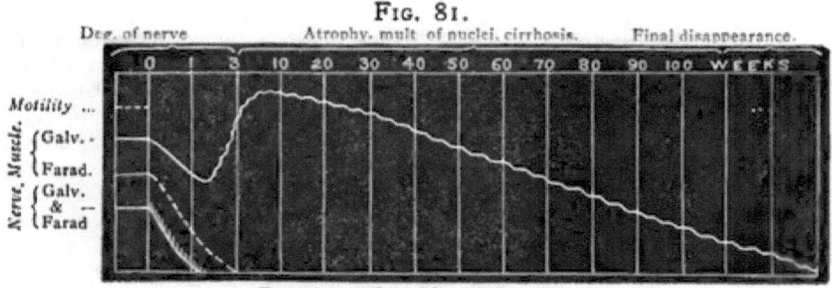

Fig. 81.

Diagram C. No recovery.

The time over which the phenomena spread themselves is indicated by the numbered ordinates (1, 2, &c., which denote weeks). The first ordinate (0) indicates the starting-point of the disease when very acute, or the moment of nerve section or injury. The dotted horizontal line indicates the voluntary power or motility, which is suddenly lost at 0, and reappears at a sooner or later stage in cases of recovery (○ in diagrams). The galvano- and the farado-muscular excitability° are denoted by two separate lines; the electro-nervous excitability by a third. The rise and fall of the several curves show the quantitative rise and fall of the reactions they represent. The wavy part of the galvano-muscular curve shows the period of qualitative alterations in the contractions. At the top of the diagram are indicated the approximate periods occupied by the various histological changes through which nerve and muscle pass during the process of degeneration and regeneration.

During the first two weeks degeneration of the nerve occurs, and as a consequence its excitability vanishes. The disappearance of farado-muscular contractility denotes the degeneration of the intra-muscular nerve elements. The first symptom of regeneration is the apparition of motility; close upon its heels usually follow the first signs of restored electro-nervous and farado-muscular excitability. The degenerative process in muscular fibre is accompanied with a quantitative (increase) and qualitative modification of its galvano-reactions. Cirrhosis coincides with a fall of the exalted excitability. The return of the normal mode and series of contractions is very gradual.

These diagrams are intended to represent typical cases of RD, and as such will rarely be found to tally in all their details with the exact course of events in an actual case. The conditions under which they are best realised are those in which the trophic centres or paths are suddenly destroyed (traumatic lesions, facial paralysis, &c.) When a more gradual morbid process is the cause of the

---

* We have already mentioned that muscular substance, even when healthy, does not react to stimuli of very short duration, such as faradic shocks. Hence there is, absolutely speaking, no such thing as farado-muscular excitability. The term is a convenient one, however, and we use it to mean the indirect excitability of muscles to faradisation applied to the nerve endings and filaments they contain, as distinguished from the same excitability to faradisation applied to their motor nerve (farado-nervous excitability).

electrical changes, the intermixture of effects due to the simultaneous excitation of fibres, some of which are healthy, others more or less altered in their structure, masks or distorts the degenerative reactions. Generally speaking, the rate of progress and depth of the lesion, as well as probably numerous other causes still hidden from us, tend to modify the phenomena in various ways. Frequently enough, the picture of RD presented to us, is but fragmentary or even fallacious; much experience and skill in conducting the experiments, can alone enable the observer to recognise the unity of type, between such instances and those schematised in the diagrams. It would be beyond the scope of an elementary work, to go into minute details concerning the numerous departures from the typical course of events just described. The beginner would only be confused thereby, instead of stimulated in his endeavours to familiarise himself with the fundamental facts of electro-diagnosis.

It is the chronological distribution and the relative intensity of the several factors combining in the production of the reaction of degeneration, which vary most considerably according to the acuteness and general anatomical and clinical characters of the lesion giving rise to them. Hence the *presence of qualitative alterations* in the muscular reactions is the main point to be attended to; for by itself it is sufficient to demonstrate the existence of the arrest or impairment of the trophic influence of the cord on the nerves and muscles.

There is one variety of RD which requires special mention here however, because it throws some light upon this "trophic" function, and is of prognostic importance. Erb has shown that in some cases the muscles alone display the typical quantitative and qualitative modifications of galvano-reactions just described, whilst their farado-excitability is preserved and the nerve continues to respond normally or subnormally to both currents. The diagram (fig. 82) embodies these facts in their chronological order, and in their relations to the histological changes upon which they depend.

FIG. 82.

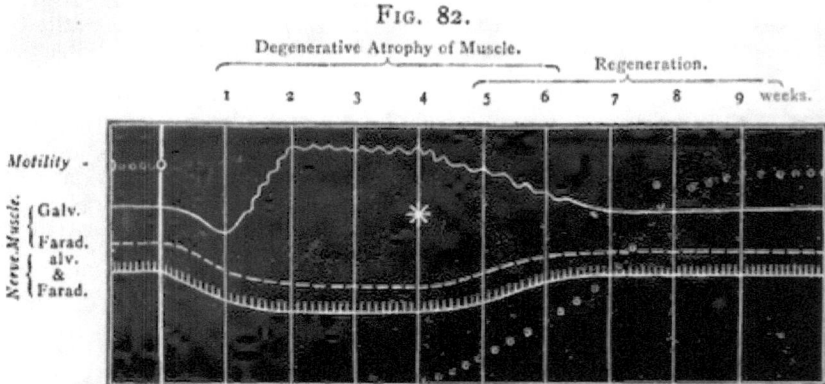

DIAGRAM D.

In this *partial* RD the nerve apparently remains intact, whilst the muscles go rapidly through a process of degeneration and regeneration. This is found in many cases of facial paralysis, where probably the pressure upon, or morbid process in, the nerve-fibres is not sufficient to impair their vitality, though sufficient to arrest their function—viz., the conduction to the muscle of trophic and motor influence. It is in certain cases of partial RD, that we find the modal alterations in the muscular response to nerve-excitation (viz., a sluggish, protracted contraction to single stimuli, galvanic or faradic) previously mentioned. The fact of partial RD occurring also in certain diseases affecting the multipolar ganglionic cells, compels us to assume a distinction between the neuro- and myo-trophic agencies. Whether this distinction depends upon special mechanisms or not, is as yet impossible to say; still in order to represent the phenomena graphically, it is necessary to assume such special mechanisms; but let it be understood that the physiological independence of the trophic influences by no means postulates the existence of special centres and fibres. Indeed the existence of any such structures may fairly be questioned. But trophic functions may perfectly well be ascribed to the motor centres, and all that is required for our argument is the recognition of the facts: (1) that trophic changes occuring in muscle are independent of paralytic phenomena; (2) that such trophic changes may occur in both muscle and nerve, or in muscle only.

The appended diagram, also by Erb, will be found convenient as embodying in a schematic manner the relations between the motor and assumed trophic centres on the one hand, and the nervous and muscular fibres on the other.

The point of convergence of the trophic, as well as of the volun-

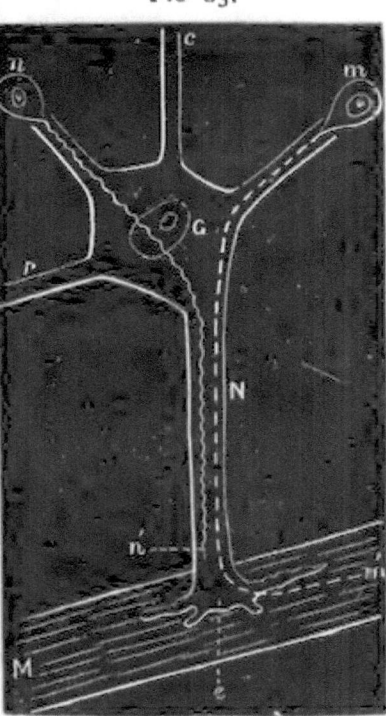

Fig 83.

Diagram E.

M. Muscular fibre.
N. Nerve-fibre, with its ending *e*.
G. Multipolar ganglion-cell, from the anterior horn of grey matter, or bulbar nuclei.
c. Path of impulse from the brain: antero-lateral columns.
r. Path of reflex excitation from the sensory sphere.
m. "Trophic centre" for the muscle.
n. "Trophic centre" for the nerve.
m-m'. Path of trophic influence to the muscle.
n-n'. Path of trophic influence to the nerve.

tary and reflex influences, is supposed to be in the multipolar ganglion-cell, G.

Thus if the lesion be at c, or what comes to the same thing in the cerebral centres above, we have as characteristic symptoms, paralysis (loss of motility), but no atrophy, and no loss of electro-excitability of either nerve or muscle (cerebral hæmorrhage, lateral sclerosis, &c.).

If the lesion involves both c and m we have paralysis with atrophy of muscle, but not of nerve. Electrical reactions: qualitative and quantitative changes in galvano-muscular irritability; the nerve responds to both galvanism and faradism (amyotrophic lateral sclerosis).

The lesion is confined to the trophic centre (m) of the muscle: the latter atrophies, but is not paralysed, and presents the qualitative alterations of reaction of degeneration; the quantitative changes are only in the direction of a diminution unless the morbid process be acute. Electro-nervous excitability present. These phenomena (partial RD) occur in progressive muscular atrophy, bulbar paralysis, mild acute poliomyelitis, &c.

When G is destroyed there is loss of motility, loss of reflex action, atrophy both of nerve and muscle. The electrical phenomena are those of the reaction of degeneration fully developed. The same phenomena must obviously be present when N is completely (*i.e.*, including m' and n') destroyed at one point. G is destroyed in anterior poliomyelitis (infantile paralysis); N in severe peripheral paralyses, whether from traumatic, "rheumatic," or any other cause.

There are, however, many instances of peripheral paralysis, where the loss of voluntary motion exists, and yet the other symptoms are not those of the "severe" forms just mentioned; they may be conveniently classed under the heads of the "light" and the "middle" form.

In the light form, paralysis is the *only* symptom. All the reactions are perfectly normal, we must then assume that N is interfered with only in its conductivity for motor impulse—M' and N' remaining intact.

In the middle form there is atrophy of the muscles and qualitative changes in their galvano-contractility; electro-excitability is present (partial RD). Here there is in addition a lesion of the path m-m'—otherwise, an interference with the myotrophic function of the nerve fibres.

In lead palsy, which is characterised by atrophic changes in nerve and muscle, the reaction of degeneration is fully developed.

The accompanying table must be taken as roughly schematic only. It is meant to fix the leading principles of electro-diagnosis in the memory; not to act as a key to diagnosis. In progressive amyotrophic diseases, for instance, the typical reactions are rarely met with. The tendency of the lesion is to extend beyond its theoretical seat. Hence partial RD is here soon masked by considerable diminution in the excitability of both nerve and muscle.

Synoptical Table, showing the connection between certain pathological states and the electro-diagnostic phenomena accompanying them.

Seat of Lesion.	Prominent Symptoms.	Electrical Excitability.	Pathological Conditions where found.
c	Paralysis. No muscular degeneration.	All normal.	Lateral sclerosis (idiopathic or from cerebral disease).
c & m	Paralysis. Muscular degeneration.	*Nerve:* Normal. *Muscle:* Qualitative and quantitative alterations (partial RD).	Amyotrophic lateral sclerosis.
m- extending to G	At first no paralysis; muscular (later nervous) degeneration.	*Nerve:* Normal, later diminished. *Muscle:* Qualitative and quantitative alterations (partial RD).	Progressive muscular atrophy (of central origin). Bulbar paralysis. Mild acute poliomyelitis.
G	Paralysis. Atrophy of muscles and nerves. Abolition of reflex actions.	*Nerve:* } RD *Muscle:* }	Poliomyelitis anterior. (Infantile or spinal paralysis). Lead poisoning.
N	Paralysis. No degeneration.	All normal.	Light form of "rheumatic," traumatic or pressure paralysis.
N & m'	Paralysis. Muscular degeneration.	*Nerve:* Normal. *Muscle:* Qualitative and quantitative alterations (partial RD).	Middle form of ditto.
N m' & n'	Paralysis. Muscular and nervous degeneration.	*Nerve:* } RD *Muscle:* }	Severe form of ditto.
M	Pseudo-paresis. Simple atrophy.	Normal, or diminution to maximal excitations.	Muscular wasting in phthisis, &c.; and in diseases of joints. Idiopathic myositis?

We are now in possession of the facts revealed by the electrical investigation of the nerves and muscles in a number of morbid conditions, and we have learnt the principles of interpretation of these facts. We know also that normal reactions depend upon a normal nutrition of those tissues, and that certain quantitative and qualitative departures from the standard correspond to certain histological processes set up when the trophic influence of the cord is in abeyance. We have insisted on the fundamental principle that there is absolutely no correspondence between the nature and amount of reaction and the degree of motor disturbance. Muscles and nerves which are hopelessly paralysed may continue for years to yield perfectly normal responses to either current, whilst reactions indicative of degenerative processes are consistent with a considerable amount of motor power. The following remarks are intended to define still more clearly the scope and limits of electro-diagnosis.

The electrical exploration of muscles and nerves does not give us any direct information as to the nature of the lesion producing the motor or trophic symptoms. The latter may be inflammatory, or traumatic, or toxic. The determination of this point rests upon the history and general clinical aspects of the case, and the teachings of morbid anatomy in parallel instances. With reference to the localisation of the process, however, much light may be thrown upon the problem, especially when RD is present. The latter allows us to conclude at once to a disease of the anterior cornua, or of the peripheral nerves; and a more exact investigation of the distribution of the degenerative changes among the muscles enables us to localise still more precisely the seat and extent of the morbid process.

When the trunk of a nerve is diseased the degenerative processes extent over and are limited to, all the muscles supplied by that nerve. When, on the other hand, the lesion affects the anterior cornua of the cord, these processes are distributed among groups of muscles having a physiological, and not merely an anatomical, connexion. The reason of this fact is that the ganglionic motor cells in the cord are arranged in groups governing the simpler coördinated movements. Hence, degenerative changes will be distributed among muscles, according to the seat and extent of the cornual lesion, and conversely the localisation of the peripheral changes gives us a clue to the position of the mischief in the cord. Such a localisation is often impossible without the assistance of a careful electrical diagnosis; as for instance, when the disease is of slow growth, and obvious atrophy and paralysis are not present. Every spinal motor root arises from a spinal segment containing circumscribed aggregation of ganglionic cells, which experimental physiology shows to govern the coördinated action of a certain number of muscles. Thus by faradisation of the several cervical, lumbar, and sacral roots in the monkey Ferrier showed that in each case a certain definite movement of the arm or leg ensued. By analysing these movements he was able to define which muscles were thrown into activity in each case, and by reasoning one step further to determine the spinal motor centres of those muscles. His results confirmed in the main the views propounded by Remak and Erb on the grounds of clinical observation, and may be provisionally, at least, tabulated as follows (see "Brain," vol. iv., 1881, p. 226):—

4th cerv.	..	Delt., rhomb., spinati, biceps, brach. ant., sup. long., ext. hand.
5th ,,	..	Delt. (clavic.), biceps, brach., serratus m., sup. long., ext. hand.
6th ,,	..	Latis. dors., pect. m., serr. m., pronators, triceps.
7th ,,	..	Teres m., lat. dors., subsc., pect. m., flex. hand, triceps.
8th ,,	..	Flexors, wrist and fingers, muscles of hand., ext. wrist and fingers, triceps.

1st dors.	..	Muscles of hand (thenar, hypothenar, interossei).
3rd lumb.	..	Ileo-psoas, sart., abductors, extensor curis.
4th ,,	..	Ext. fem. et cruris, peron. long., abductors.
5th ,,	..	Flex. and ext. toes, tibial, sural, and peroneal muscles, ext. rot. thigh, hamstrings.
1st sacral	..	Calf, hamstrings, long flexor big toe, intrinsic muscles of foot.
2nd ,,	..	Intrinsic muscles of foot.

We see thus how from the occurrence of RD in certain groups of physiologically correlated muscles we can infer the spinal seat of the morbid process upon which the peripheral condition depends.

Electroprognosis.—It is obvious from the fact that qualitative departures from the normal polar formula depend upon histological changes in the nerves and muscles, that electrodiagnosis gives us a clue to the depth of the lesion, and its rate of progress; hence it becomes in some cases a valuable means of framing a prognosis. Erb has shown how in cases of facial paralysis for instance the duration of the process of degeneration and regeneration could be accurately determined by the behaviour of the electrical reactions; we reserve this point for a fuller discussion hereafter. Generally speaking, the prognosis becomes gloomy in proportion to the persistence of the RD beyond the usual period of regeneration (see diagrams) but it may not be unnecessary to guard the student against the error of looking upon the occurrence of alterations in the response of nerves and muscles as in itself indicative of irreparable mischief. On the contrary, RD is often of far more favourable prognosis than normal reactions, which we have previously found to be consistent with absolutely incurable lesions, involving complete paralysis, as well as with intractable spasms, tremors, or convulsions. Perfect restoration to health, on the other hand, is often witnessed after the deep nutritive changes of nerves and muscles in paralyses of poliomyelitic or peripheral origin.

THE COMPARISON OF EXCITATIONS.

Having learnt the facts of electrodiagnosis, and how under certain conditions the reactions of nerves and muscles present quantitative and qualitative alterations, we come to the discussion of the methods of ascertaining these facts.

The presence and degree of an alteration can be determined only by comparison. Now, every experiment involving a comparison, in order to yield reliable results, must be made under definite conditions. These conditions, in the two instances compared, must be made identical, save upon the particular point which it is the object of the experiment to investigate.

In testing a nerve with a view to determining quantitatively its excitability, we must ensure that all the conditions under which the experiment is performed be precisely the same as those under which our standard of comparison has been obtained, or at least bear to them certain known relations. These conditions will be presently enumerated and discussed.

Qualitative alterations of the modal category, involve comparison not of quantity but of character. Their determination is easier and less liable to fallacies, because the experimental procedure by which we obtain it, requires no fine quantitative adjustments; on the other hand, a much more practised eye is required to recognise the finer gradations of an altered muscular curve. Modal alterations are recognised by the more sluggish and less complete character of the muscular contraction, as compared with what we have learnt in the physiological part to be the normal contraction. The myographic curve of the one presents a less steep ascent, a more protracted summit, and a more gradual descent than the curve of the other.

Serial alterations consist in a disturbance of the order in which the polar closure and opening reactions occur under conditions of increasing current strength. The comparison is one of relations; we have not to compare the absolute exact quantities of electricity required to bring about each of the members of the polar formula in the explored nerve or muscle, with the quantities required for the same purpose in the standard organ; but to determine whether the proportions which these quantities bear to one another in the one series, are the same as those obtaining in the other series—in other words whether the several kathodic and anodic closure and opening contractions make their appearance to current strengths growing in the same proportion in the two series. In testing for qualitative serial alterations of the formula, therefore, we are not subjected to the difficulties arising from the necessity of localising equal excitations in two different structures; it is enough if we localise two series of excitations in corresponding points of the body, and note in each case the strength of the current necessary to elicit the several reactions.

Owing to the variations in the resistance of the skin in different individuals and at different points of the body of the same individual, and to the variable relation of different nerves to the skin and the underlying tissues, it is obvious that the only trustworthy standard with which we may compare any reaction of a diseased nerve (or muscle) is the same reaction obtained on the symmetrical healthy organ on the opposite side of the body. The method of effecting the comparison will be set forth presently.

Whenever a nerve presents abnormalities on both sides of the body, it becomes necessary to have recourse to a less reliable standard; and this may usually be obtained by the exploration of some other nerve which experience has shown to yield, on the average individual, minimal contractions to a strength of current, either similar to, or at least differing by a known proportion from, that required to excite the explored nerve. When, for instance, one of the peroneal nerves has to be tested, and its reactions cannot be compared to those obtained on the other peroneal, recourse should be had to the ulnar for instance, or to the spinal accessory, or to the facial, which experience shows to react to currents not widely different in strength from those necessary to excite the peroneals in the average specimen of mankind.*

The comparison of the excitability of two different nerves is always a task of some delicacy; but when in addition, the subject of the experiment is very fat, or presents other such peculiarities, the demonstration of quantitative departures from the normal excitability, becomes a matter of great difficulty and uncertainty.

Everyone therefore who is called upon to investigate the reactions, under conditions of bilateral disease, should be familiar with the scale of excitability[†] of certain conveniently chosen nerves in the healthy subject.

As a means of obtaining this familiarity, and as an exercise to be sedulously practised on numerous healthy individuals, preparatory to investigating diseased subjects, the following experiment is here described:—

The positive pole of the faradic current being connected to a large electrode held steadily upon the sternum or back of the subject, the negative is connected with a fine electrode, fixed to an interrupting handle. The observer then seeks the weakest current (expressed in terms of the distance between the primary and secondary coil) necessary to obtain a muscular contraction, when the electrode is applied to the temporo-facial (branch to frontal muscle), the spinal accessory, the ulnar, the peroneal nerves. The spots at which the excitation is made on each side of the body must be exactly symmetrical. Having made a table of the numbers so obtained the observer proceeds to test the resistance of the body at the places where the current has been applied. For this purpose he connects the large electrode (which remains unmoved on the trunk) with the positive pole of a galvanic battery of which 10 cells are put, with a galvanometer, into the circuit; and exchanging the fine electrode for a middle-sized one, connects the latter with the negative pole and applies it successively over each of the nerves which have been excited. He reads on the galvanometer the deflections obtained in each case and writes them down opposite the corresponding numbers of the table.[‡]

Every time the electrode is applied to the skin, whether for exciting or estimating the resistance, it must be moistened afresh.

A concrete example will be useful to make this description clear, and show more fully the bearings of the results obtained. The following numbers were obtained in an experiment:—

---

* This is the method devised by Erb and published by him in a paper on "Tetany". (*Arch. für Psych.*, 1873. Cf. "Elektrotherapie." Lecture viii).

† It is not meant here that the various nerves differ in their *absolute* excitability, but only in the readiness with which they are reached by percutaneous excitations, under varying conditions of depth, environment, etc.

‡ The reader who has mastered the first chapter of this book need not be more than reminded that the galvanometric deflection indicates the strength of the current given in each case by the 10 cells, and is therefore in each case inversely proprotional to the resistance of the part of the body included in the circuit. (Ohm's law: $C = \dfrac{E}{R}$).

Nerve excited.	Distance of Coils (in millimetres) at which first contraction appeared.		Approximate Galvanometric deflection with 10 cells.	
Temporo-facial	Right 130	Left 132	Right 6·1	Left 6·3
Spinal accessory	,, 124	,, 123	,, 4·8	,, 4·8
Ulnar	,, 119	,, 118	,, 4·4	,, 4·3
Peroneal	,, 121	,, 119	,, 3·8	,, 3·5

The experiment is completed, if necessary, by finding the strength of galvanic current requisite to produce a contraction, when the circuit is closed, with the negative pole on the nerve (KCC), in each of the 8 cases tabulated above; and the number of cells, or better still the galvanometric deflection, written down by the side of the corresponding reading of the induction scale.

It will be found that on every healthy subject there is a certain proportion preserved between the numbers forming each series of the above table. Variations occur owing to the extraordinary complexity of the conditions governing the results; but personal experience will teach that those variations are not considerable, and, as we shall have the opportunity of showing more fully further on, that they often are explained by differences in the resistance of the skin or tissues at the points of application. The recorded galvanometric control gives a clue to such differences, and serves to eliminate the fallacies arising from them.

The table shows us that in the typical individual we tested, the two nerves of each pair reacted to much the same current, but that the temporo-facial and spinal accessory responded to apparently weaker currents than the ulnar and peroneal; we say *apparently* because, as the galvanometer shows us, the resistance of the tissues in the case of the face and neck is less than in that of the arm or leg; and because the numbers taken from the induction-scale represent electromotive force,° not current strength.

The practice of such observations as these, is to the electro-therapeutist, what the study of healthy heart and lung sounds are to the stethoscopist. It is essential to the framing of a standard of comparison, without which any attempt at estimating departures from the normal must necessarily prove futile.

THE CONDITIONS OF EQUAL EXCITATIONS.

When two effects are to be quantitatively compared it is essential that the amount of force used as reagent should be precisely similar in each case. Now the cause of the muscular contractions we are studying is a certain alteration in a definite portion of irritable

---

° We have mentioned (p. 74) that the electromotive force of a coil increases much more rapidly than the arithmetical ratio adopted in the graduation of the scale (in millimetres, etc.). Thus in the example given above the coil at 120 mm. might very well be supposed to give an electromotive force almost double of that obtained when it stands at 130 mm.

CONDITIONS OF EQUAL EXCITATIONS.

tissue by the passage of an electrical current; and this alteration is proportional to the *quantity* of electricity conveyed by the current through that portion, in other words to the electrical density in those tissues.

From the principles illustrated in the former part of this book we know that the electrical density in the embedded nerve over which one electrode is applied is the product of four factors:

1. The absolute strength of the current—(itself determined by the electromotive force and the resistance in circuit).
2. The size of the exciting electrode.
3. The relative position of the exciting electrode, with reference to the underlying nerve.
4. The specific resistance of the surrounding tissues, with reference to that of the nerve.

But we have already had the opportunity of convincing ourselves that a contraction is proportional not only to the degree of electrical alteration to which a nerve or muscle is subjected, but also to the rate at which the change is developed (see Electrophysiology, Exp. III). We must therefore ensure a perfect equality of the *time* taken by the current to reach its maximum at each make, or closure of the circuit, and its minimum (*i.e.* to cease) at each break or opening. This can be done only by making and breaking the current in the metallic part of the circuit with a properly constructed mechanism* whilst the electrode remains immoveably applied over the nerve or muscle examined. "Dabbing" the skin with the electrode, as is often done, is to be rejected for the quantitative estimation of excitability on account of the irregularity of the makes and breaks thereby produced, as well as of the inevitable shifting of the electrode at every make from the exact spot chosen over the given nerve.

When speaking of the "motor points" we had an opportunity of showing that the excitability of the muscular surfaces varies greatly from place to place according to the presence or absence of motor nerve filaments. No two contractions are comparable which have not been obtained by localising the stimulus in portions of muscles similar in this respect (not for any physiological, but for the physical reasons emunerated above) in the intensity and nature of the effects produced. Hence a condition to be fulfilled with reference to the *place* at which comparative experiments are to be made. The portions of tissue in which the electrical change is effected must be similar in every case.

The *quantity* and *time* conditions, constitute the factors of Du Bois-Reymond's law; for which we may for our purpose (including the *place* condition) substitute the fundamental axiom: "that in order to obtain absolutely comparable results we must ensure that equal quantities of electricity vary, in equal and similarly placed portions of contractile tissue, in equal times." Now it is impossible with the

---

* Many interruptors are so badly constructed as to give a series of excitations at make and at break. Instrument makers should be impressed with the importance of attending to apparently insignificant details of construction—here as in numerous other instances.

most accurate instruments, and the most skilled use of them, to secure fully these conditions in the human body. We can, indeed, by means of the galvanometer, control the quantity of electricity flowing in and out at the electrodes; and, by means of an interruptor, make and break the current with a uniform rapidity. But beyond this we can only approximately carry out the rule. By endeavouring to secure an accurate localisation of the electrodes, and a steady pressure upon them, we diminish the chances of an unequal diffusion of the current in the deeper structures to be influenced. This diffusion, however, further depends upon conditions beyond our ken, that is, upon the state of the deep tissues themselves, as affecting their relative conductivity, which in health we may assume to be more or less constant—but not so in disease. Here, therefore, we have a source of fallacies which no precautions can remove, and which are occasionally sufficient to vitiate the most carefully conducted experiment. Practice and experience alone can teach us when there is reason to believe that a departure from the normal type of contraction is partially or entirely due to physical, instead of physiological or pathological causes.

THE GALVANOMETER IN DIAGNOSIS.

We have spoken of the galvanometer as the means of eliminating one of the fallacies—the leading fallacy—besetting electrodiagnostic investigations. It is strange indeed to hear men who ought to know better gravely assert the presence of increased or diminished irritability in a nerve or muscle without any reference to the resistance of the body in the particular instance. If the galvanometer is mentioned to them they object that their purpose is a purely practical one; they do not want to measure anything, but only to estimate the reactions of a diseased organ; they say that the needle takes too long a time to settle; and that after all the galvanometer is applicable to the galvanic current only—the faradic cannot be so measured. These objections betray a complete misconception as to the application of the galvanometer to electrodiagnosis. Actual measurement of current, it is true, can be dispensed with in the practice of diagnosis but not so the elimination of fallacies due to variations in the resistance of parts of the body included in the circuit. The galvanometer has to be used in electrodiagnosis for the purpose of fulfilling this condition. What should we say of a physician who after examining each side of the chest with a stethoscope of different conducting power, stated that the sounds were louder on one side than the other, and thence concluded to an abnormal condition of the lungs? And yet this is precisely the mistake to which those who reject the galvanometer as useless in diagnosis are constantly exposed, and which makes their statements concerning the augmentation or diminution of excitability in numerous cases or kinds of disease of no value. The method of eliminating the influence of variable resistances consists in taking a convenient number of cells, say 10, and registering the deflections given when the electrodes are applied over the nerves and muscles which are the subject of the electrical test on either side of the body. The numbers so obtained are inversely proportional to the resistances of the several points tested, and reveal whether any apparent difference in the reactions arises from a difference between these resistances, (see p. 137). The following example will make this point clear, and serve as an example for practice.

On examining the reactions to faradism of several nerves of a man on the two sides of the body, the following numbers were obtained. They express (in millimetres) the distance of the secondary

	Distance of Coils.		Deflection of the needle in mw. 15 cells.	
Facial	Right 224	Left 228	Right 9·4	Left 9·3
Accessory	,, 236	,, 239	,, 10·1	,, 10·2
Ulnar	,, 212	,, 190	,, 4·3	,, 3·6
Peroneal	,, 215	,, 218	,, 4·5	,, 4·2

coil from the primary when the first contraction is observed.

The usual method was used, viz., a large electrode was fixed to the trunk; the other (small) was applied over the several nerves mentioned, both were well moistened. What strikes us at once is the disparity between the apparent strength of current necessary to excite the right and left ulnar respectively. Had the case been one of alleged railway injury, had the man complained of subjective sensations or muscular weakness, there would have been no dearth of witnesses ready to swear, after an examination conducted according to the popular notion, to an organic disease of some part of the brain, cord, or nerve, or all three. But in this case the man was perfectly healthy.

The apparent paradox vanished the moment we took the precaution, after the faradic examination, and before drawing any conclusions from the results it yielded, of consulting the galvanometric needle. The revelation obtained thereby of a notable difference in the resistance of the skin or tissues on the two sides of the body, showed that the increase in electromotive force which had to be called into play before we could obtain contractions on the right side, was necessary in order that the testing current should be brought to the same strength as on the left side. The practical lesson derived from this and other similar instances is obvious; it may be added that if the danger of mistaking variations in the tissue resistance for variations in excitability may be so great in unilateral disease, it becomes greater still where, both sides of the body showing apparent abnormalities, we have no standard by which to determine how far the departures from the normal are caused by alterations in the conductivity, rather than in the excitability of the tissues. In all such cases the use of "high tension" currents (see page 30) is of material assistance in eliminating the influence of increased tissue resistance.

OTHER FALLACIES OF DIAGNOSIS.

We need not say much with reference to the physiological sources of fallacy in electrodiagnosis. The alterations in the excitability of a nerve due to previous polarisation or excitation might, under certain circumstances, lead to conflicting or misleading results. When the effects of the two poles are successively tested it may easily happen that the contractions are exaggerated to that pole which is used after the other. In order to eliminate the influence of after effects, we must abstain from exposing the nerve or muscle to long continued galvanic influence, and excite at intervals sufficiently long to allow the disturbing effect of previous excitations to pass off. Again, when the opening contractions are investigated we must adopt a definite rhythm for the making and breaking of the current, owing to the influence of the duration of the current upon the OC. In all cases of doubt repetition of the examination on the rested nerve or muscle becomes necessary.

There are a number of adventitious circumstances connected with the testing of nerves and muscles in the living body, which tend to obscure or distort the evidence of abnormal reaction. The chief of these is the diffusion of the stimulus to neighbouring parts, and the consequent admixture of muscular effects. This is especially the case when the excitability of the organ tested being diminished, strong currents are required to excite them. We then find that the whole limb or part is thrown into action, and it becomes necessary carefully to distinguish whether there is any contraction of the muscle under observation. Atrophy of the explored muscles also favours diffusion of the current into adjacent parts. If for instance we have to test a case of saturnine wrist drop, we shall usually find that not only the neighbouring extensor muscles, but that the flexors also, partake in the movements. The contractions of associated muscles tend to hide the real deficiency in the action of the diseased ones; whilst the contractions of the antagonists tend to counteract and efface the effect produced by the contraction of the muscles under examination. Such an intermixture of effects cannot always be prevented, and we must keep a vigilant look out for the amount of movement observed at the surface, or at the tendons, of the diseased muscles. It is always good to keep the limb in a such a position as to give the freest possible play of these muscles, whilst the associated movements are kept in check by appropriate mechanical means.

A still more difficult task is that of distinguishing between the reactions of degenerating bundles of muscular fibres, and the healthier tissue with which they often are surrounded. In several forms of degenerative atrophy the whole muscle is not affected "en masse" but piecemeal. We then find that the maximal contractions are diminished, though the minimal may be obtained with low current strengths. Though the kathodic and anodic closure contractions may apparently be serially normal, there may yet be observed, following the latter chiefly, typical sluggish contractions of degenerating fibres.° It is the want of attending to this fact which makes many observers miss the presence of degenerative reaction in progressive muscular atrophy and other chronic atrophies of "spinal" origin.

THE PRACTICE OF ELECTRODIAGNOSIS.

The thorough and accurate investigation of a man's electrical reactions is, we repeat, a physiological experiment of a very delicate and complex nature, very different indeed from what commonly passes under the name of an electrical examination. It requires a knowledge of the principles, physical and physiological, laid down in the previous chapters of this book, and in addition reliable instruments and skill in using them. A few words concerning the method of carrying out an actual investigation, will therefore not be superfluous.

The subject must be so placed that the symmetrical body be equally accessible to the operator; the light must be good and side shadows avoided. It is very difficult indeed to examine a patient in bed, owing to the difficulty of getting readily at the symmetrical points of the body. The muscles to be excited must be in an equal state of relaxation, and the patient instructed neither to assist nor resist the movements imparted to the limb by the excitation.

One electrode of large size (a plate of flexible metal) is fixed securely to the trunk, and this is by no means such an easy process as one would think. The sternum is preferred as a point of application by the highest authorities. In some cases the patient can keep it fixed himself with one hand; or it may be secured by a broad elastic band round the body, or by the dress. In order to ensure uniform pressure upon the plate, it is desirable to stuff in between the plate and hand or dress an artfully crumpled napkin. These details, insignificant as they look, are essential to the smooth and successful carrying out of the experiment.

The other electrode, of smaller size and fixed to an interrupting handle, is held in the operator's hand.

In performing all electrodiagnostic operations care must be taken to use water, or salt and water, plentifully; and not to allow the current to flow longer than is absolutely necessary, for each excitation, or when the galvanometer has to be consulted, for the needle to come to rest. In the examination of the nerve trunks and motor points, it is best to begin with the *faradic* current. For this purpose the "fine" or the "small" electrode is to be used, ac-

---

* I have had under observation a case of progressive muscular atrophy in which such a phenomenon, but on a larger scale, was beautifully developed. Galvanic stimulation of the flexor muscles of the forearm, was followed by a distinctly double contraction as shown by the movements of the tendons at the wrist. Whilst some reacted fairly vigorously and rapidly, others displayed the sluggish type of contraction and were seen to rise and sink after the others had accomplished their jerk.

cording to the size of the structure to be excited. The main conditions to be fulfilled are the accurate localisation of the current upon symmetrical points, and the thorough moistening of the electrode and skin. As the operation has often to be repeated several times, it is convenient to mark these points, when found, with a copying-ink pencil. We determine at what distance of the secondary from the primary coil, the muscles begin to react. If the excitability is low, strong currents have to be used; it is then necessary, on account of the pain, to make slow interruptions by moving the hammer with the finger, or by a spring rheotome in the primary circuit. When the differences are slight, repeated examinations are imperative.

The previous remarks referring to the intermixture of muscular effects must be kept in mind, and the necessary precautions taken to eliminate them—here and throughout the examination. The electrode must be dipped in the liquid every time it is applied to the body. Having determined the minimal excitability of the nerves and muscles, we complete our examination, if desirable, by estimating the amount of contraction obtained with a current of medium strength; and determine also the maximal contraction obtainable with very strong currents. When the amplitudes of two muscular contractions have to be compared, it is advantageous to use a bifurcated rheophore with two equal interrupting electrodes, which are held, one in each hand, over the corresponding points in the two sides of the body. The current is then sent alternately through the one and the other; any difference in the amount of movement produced is far more readily observed thus, than when a single electrode is transferred from one side to the other.

We next take a galvanic current of 10 cells, with the galvanometer in circuit, and note the deflections of the needle when the electrode (of medium size) is placed over the points to which the testing current has been applied (see page 137). The deflections betray any differences which may exist in the resistance between the two sides; and if we are using instruments familiar to us, enable us to measure these resistances and compare them with others if we choose to do so.

We finally proceed to the examination by means of the *galvanic* current. The electrodes are to be larger than those used for the faradic current (the "small" disk about ¾ inch in diameter (p. 90) answers very well). The nerves and motor points are excited by makes and breaks of the current. We note the number of cells (or better still, if we have a dead-beat galvanometer,° the deflection) at which the various reactions appear. We begin with the kathode then proceed to the anode, making and breaking the current with equal periods of flow. It is here that the influence of electrotonic after-effects have to be guarded against. The current must not be allowed to flow long, and kathodic and anodic excitations made to alternate several times in succession, allowing an interval to elapse

---

° *i.e.* a galvanometer in which the swinging of the needle is diminished by some artifice of construction.

between each change of pole. Having thus established the "polar formula" of the nerve examined, that is to say observed the relative current-strengths at which the several closure and opening contractions occur when each pole is on the nerve (KCC, ACC, AOC, KOC), and at which the closure contractions become tetanic during the period of flow (KDT, ADT), we proceed to investigate the reactions of the muscle themselves to galvanic excitations.

Taking an interrupting handle fitted with a "small" or a "medium" electrode according to the size of the muscle, we proceed to examine the reactions to the closure and opening of the current, when the kathode and when the anode is on the muscle observing the same general rules of procedure as in the examination of nerves. We have said previously that qualitative alterations (serial or modal) are not usually met with in the galvanic examination of nerves. In that of muscles, on the contrary, they form the most important part of the process—though quantitative abnormalities are by no means to be overlooked.

The normal polar formula of muscle is not so complex as that of nerve. It is difficult to obtain by the direct galvanic excitation of a muscle more than two reactions, viz., KCC and ACC. Whenever an opening contraction or a duration tetanus is readily obtainable we may assume an abnormal constitution of the muscular fibre. Another peculiarity of normal muscular reactions is, that owing to the complex structure of the organs which contain a variable intermixture of nerve elements, there is a certain relation between the current strengths necessary to produce KCC and ACC, and it is somewhat variable. It is even possible on perfectly healthy muscle to obtain the ACC before the KCC. The explanation of these facts is to be found in the fact that the excitation falls in the kathodic (polar or peripolar) zone of polar action; according therefore to the presence or absence of the more excitable nerve elements, in one of the two zones, the effect of the positive or of the negative pole will be enhanced. It is important therefore to remember this source of fallacy in testing muscle for serial alterations, though its practical importance is less than appears at first sight. The reactions of a muscle at one point of its surface must however not be compared to those at another point of the same muscle, but to those at the corresponding spot of the same muscle on the other side of the body or on some other person.

The estimation of modal changes in the muscular contraction requires a well exercised eye—especially when the altered muscular fibres are mixed with healthy ones, and the mixed effects of their simultaneous excitation have to be mentally disentangled. No description can supply the place of personal experience; here as everywhere else in electrodiagnosis the one thing needful above all is Practice.

Chapter IV.

ELECTROTHERAPEUTICS.

A. GENERAL—INTRODUCTORY REMARKS.

THAT electricity is a valuable agent in the palliative or curative treatment of numerous morbid conditions is a fact now too universally recognised to require any illustrative proofs in these pages. If on one hand a large number of the reported successes of its therapeutical applications have no existence save in the imagination of the writers, yet on the other, not a few of the failures experienced in its use by otherwise able physicians are to be ascribed to the ignorance which still prevails of the rationale and methods of electrisation. The object of the present chapter is to give a condensed yet practical account of this part of our subject. But I would at the outset warn the reader that much personal practice is required before he can hope to carry out satisfactorily the apparently simple directions for the electrical treatment of the various cases in which it is indicated. Let him ever remember that he has to treat patients, not diseases; and to use an agent as multifarious in its mode of action upon the organism, as in its external manifestations.

The tendency of modern therapeutics is to become "physiological," to explain the effects of remedies on the diseased organism by their effects on the healthy organism. Now it is obvious that this tendency may be, and is frequently, pushed too far. The action of many of our most potent drugs escapes all our methods of physiological experimentation. Moreover we must not forget that diseased tissues react differently from sound tissues to the same influences. A perturbation set up in the function or nutrition of a cell or fibre which has become the seat of some abnormal process may be followed by a restoration to health; whilst the same perturbation brought to bear upon sound tissue is met and neutralised by the tendency of the normal organism to preserve its vital equilibrium. Thus the administration of the bromides may affect an epileptic brain differently from what it does a healthy brain.

Our positive knowledge concerning the physiological action of electricity is limited to its effects upon the excitability of nerve and muscle. Everything beyond these narrow limits is involved in comparative darkness. Now it is obvious that we can no more attribute the curative influence of the current upon diseased tissues to these exciting and modifying properties, than we can ascribe those of arsenic or iodine to their obvious physiological effects. We are compelled in both instances to assume that these agents set up certain nutritive changes of a deeper nature, whereby the tissue elements are enabled to resume their normal functions and constitution. Attempts have been made in the case of electricity to explain the occurrence of such changes by its well known chemical and physical, as well as by its physiological, properties. We may provisionally preserve the name of "Catalysis," given by Remak to the complex influence of the galvanic current on nutrition due to :—

1. Its property of conveying liquids from pole to pole (cataphoresis and osmosis) through the tissues, whether cell-walls, or intercellular material.

2. Its property of inducing chemical changes in solutions through which it circulates (electrolysis).

3. Its effect upon the circulation of lymph and blood through the tissues. This effect is *a.* direct (by excitation of vessels themselves), *b.* indirect (of vasomotor or sympathetic nerves), *c.* reflex (of sensory nerves).

To which list we may add—(4). Its possible excitation of the trophic influence of nerves on tissues; and of their constituents cells themselves.

The effects of the faradic current may also be defined as very slightly modifying—markedly exciting and moderately catalytical. Little is known concerning them, however, and they must necessarily vary with the method of faradisation employed.

INFLUENCE OF DIRECTION AND OF POLE.

Currents are said to have a descending or ascending direction according as they flow with or against the natural volitional impulse in motor nerves. In the opinion of some observers, it is this direction which determines the effect (exciting or depressing) obtained by electrizing a nerve. The reasons why we dissent from those who base their methods of treatment upon the alleged physiological influence of the directions of electrical currents in the body are, 1st, that in physiological experiments on human as well as on frog's nerves all the phenomena observed are best explained on the theory of polar influences only; 2nd, that it is impossible to send a current of any density in a given direction through a moderately long piece of embedded nerve; 3rd, that catalytical effects are independent (so far as we can ascertain) of the direction; 4th, that the theoretical claims set up in favour of specific indications for the use of ascending or descending currents in the treatment of various disorders are not justified by experience. In fact among the writers who have based their systems upon the assumed directional differences a great deal of divergence prevails as to the indications for the current to be used; whilst their practical results agree among themselves, as well as with the results obtained from the application of the "Polar Method" to treatment. The latter, which we saw was the only method possible in human electrophysiological and in diagnostic inquiry, consists in paying attention only to the pole to be applied to the suffering part. The effect of the other pole is said to be eliminated by placing it (as we have shown already) upon a distant, "indifferent," part of the body. The indications for the use of the anode and kathode respectively are, according to the theory, the electrotonic effects: where you want to depress nerve-activity, as in neuralgia or spasm, use the positive pole so as to produce anelectrotonus (or diminished excitability in the nerve). Where you want to excite nerve activity use the negative pole so as to produce katelectrotonus (or augmented excitability).

This theory appears very rational, but is it sound—asks the sceptical reader—is it confirmed by practical results? Several objections may be raised against it. First we know that whichever pole we apply over a nerve, we create in that nerve two virtual electrodes of opposite names, an anode and a kathode, so that we never can eliminate altogether the effect of the other pole. Next we know that excitability is diminished in the anodic region only during the short time the current actually flows; as soon as the latter is broken the nerve passes into a state of increased excitability. Again we know nothing about the intimate processes upon which the electrotonic conditions depend; nor do we know much more about the molecular state of the nerve when the seat of "neuralgia" or "paralysis." There really does not seem to be any analogy between the electrotonic states and the functional increase or diminution of nerve activity characteristic of pathological states. We have no right to infer that the production of anelectrotonus for instance, can have such an influence upon nervous nutrition as to become curative in such widely different conditions as neuralgia and spasms. Finally, we may remark that the "catalytical" influence of the current is probably of more importance therapeutically than its electrotonic effects, and we have no grounds to assume that there is any specific difference between the action of the two poles from this point of view. So far as we know the successive action of the two poles upon the tissues to be influenced in their nutrition is more energetic than that of either pole alone.

Thus we see that the "polar" method of treatment rests upon a very slender theoretical basis; does it stand the test of experience? If we take up the works of various authors we find that there is a pretty general consensus with reference to the cases where the current has done good and where it has failed. Now, as we have just said, some are in the habit of treating upon directional, others upon polar, principles; and it is easy to see that frequently enough the negative pole according to one method has to be applied where the other method indicates the positive pole, and vice versa. If then, in the majority of cases at least, either pole may be applied with equal chances of success, we do not see that the polar method has more practical than it has theoretical probability. Such are the objections which the sceptical observer might raise against the polar method of treatment, and that they are not devoid of a certain weight is shown in that Erb himself in his recent work, though a warm supporter of the polar method, fully recognises the impossibility of basing upon it a complete system of electrotherapeutics the latter must for a long time to come rest upon an empirical foundation.

I think that on the whole, the polar mode of treatment has not justified the claims of Brenner to have established a physiological method; though I am prepared to admit that there are some facts on record which prove the occasional therapeutical differentiation of the two poles. But there is another aspect of the question which is not to be overlooked. From a *physical* point of view, the polar (or rather *unipolar*)[*] method frequently offers a great advantage over the directional (or *bipolar*) method. The first condition to fulfil in electrisation is to reach the organ to be influenced with a current of sufficient density. Now, as we shall see presently, this in many cases is best effected by placing only one electrode over it, whilst the other lies at a distance.

Hence the unipolar method, in the restricted sense just given to this expression, deserves acceptance as the usual mode of applying the current for therapeutical purposes. But we may go one step further and in order to acquire methodical habits in the medical application of electricity—without which much time and labour are sure to be wasted—and in view of reaching the ultimate settlement of this and other questions, temper the rules of procedure derived from physical and empirical data with a judicious admixture of physiological hypothesis. We cannot do any harm, if we gain no signal benefit, by observing up to a certain point the polar dictum, and applying the positive pole to those parts where an anodyne or sedative effect is desired, the negative where an excitant is needed—especially if we keep in mind the purely theoretical grounds upon which we act, and know when to depart from the routine. Whenever the expected results do not follow the influence of the theoretically indicated pole, the contrary pole is to be tried. Whenever, also, the "catalytical" effects of the current appear to be promising, the alternate influence of both poles is to be exerted upon the suffering part not by sudden reversals of the current (voltaic alternatives), however, but by the gradual removing of one pole, and the gradual bringing into action of the other. In other cases the choice of the "active" pole may become a matter of convenience, as for instance when the skin is very sensitive. The positive pole may then be used, whilst the negative in the shape of a large plate, is connected with the "indifferent" electrode.

To sum up: the position of the poles in therapeutical applications is to be governed chiefly by physical principles; it must be such as to secure the most complete permeation of the organ or tissues to be influenced by the current. Whenever the unipolar method answers this purpose best, it has to be used, apart from any theoretical views of polar action.

We shall have the opportunity further on of defining the indica-

---

[*] In physiology the term "unipolar excitation" is used in a somewhat different meaning it is true. I prefer the term unipolar to that of polar because the latter implies the physiological theory which has so little to do in therapeutics; the word "unipolar" conveys nothing more than that one electrode only is applied to the part to be influenced.

tions for the use of the bipolar method; for the present the general statement will suffice, that no "directional" but only physical reasons must influence us in adopting it. When the two poles rest upon the diseased part it is obvious that the electrotonic influence of the current is out of the question; catalytical or alterative effects can alone be thought of, and then the two electrodes must be made to alternate their polarity by gradual current reversals.

CHOICE OF CURRENT.

The golden rule that "the patient has to be treated, not the disease" applies to electrical as well as to all other forms of therapeutical methods. Still there are certain general empirical principles from which we cannot depart without committing serious mistakes in the curative applications of the current. Certain general lines of treatment have to be followed, and a certain unity of procedure respected among the many modifications required in special instances. It often happens that one has to feel one's way in the search for the most effectual mode of electrisation; but experience has shown what mode answers best in the majority of cases of a certain type, in neuralgia, for instance, in atrophic paralysis, and the like. It is, therefore, only when such a mode has proved useless after a fair trial in a particular case that one is justified in resorting to other modes in the hope that they may prove successful.

Though apparently a simple task, it is sometimes far from easy to formulate definite indications as to which current, and which mode of application of either is to be preferred, in many of the various disorders amenable to electricity. The physical differences between faradism and galvanism are immense, their physiological effects very distinct, and yet their therapeutical results in certain cases equally certain. Generally speaking whenever excitation of sensory or motor nerves is required, faradism is the best means of obtaining it. Thus in the treatment of anæsthesia, or for purposes of counter-irritation, dry faradisation with the wire brush has to be resorted to. Again, when powerful contractions have to be provoked, as in certain forms of rheumatism and old joint disease, the induced current, with moist electrodes, is indicated. On the other hand in the majority of spasmodic, neuralgic conditions, and the like, galvanism is alone of any service. In atrophic paralyses the faradic is inferior to the galvanic current.

By using sufficiently powerful galvanic currents and interrupting them sufficiently rapidly we may obtain effects very similar to those of faradism. But the latter has the inherent advantage of being generated by apparatus of a much more compact and simple structure than the galvanic; so that though a galvanic battery has a far greater scope in therapeutics and might be made to do all that is ever wanted, yet on the grounds of convenience, the faradic current will always preserve its place as an agent for treatment as well as diagnosis.

The dictum "use whichever current produces the best muscular contractions" is based upon the erroneous assumption that the calling forth of muscular contractions has, *per se*, a great therapeutical virtue. It is true that as just remarked the galvanic current is of greater use in atrophic paralyses than the faradic whose exciting effect is lost in degenerating muscle. But this superiority is not so much due to the muscular contractions which may be provoked by the galvanic current in such cases, as to its deeper effects upon the circulation and nutrition of tissues. The common mistake that the good done in a case of paralysis is proportional to the amount of "artificial gymnastics" through which the muscles have been put, is strongly to be deprecated. It has often been said that such gymnastics prevent wasting of the muscles and keep them in a fit state for the time when nature has cured the primary lesion, and the impulse of the will again reaches them. Such a view is entirely opposed to our present knowledge of the matter; for in such cases where the muscle reacts normally, no wasting occurs even after years of inactivity; whereas in degenerating muscle, experience shows that no amount of galvanic stimulation prevents the downward process, which lasts until the trophic influence of the spinal grey matter again makes itself felt through the regenerated nerve. It is possible that some "artificial gymnastics" may at this period become a factor in the undoubted results obtained from electrisation under such circumstances, when, left to nature, the patient usually makes but an incomplete recovery.

In the numerous cases where no definite indication exists for one of the currents to the exclusion of the other, both faradic and galvanic currents may with advantage be used alternately, or better still conjointly as described under the heading of galvano-faradisation.

POSOLOGY.

The measurement of electrical quantities used for therapeutical purposes, in other words the dosage of electricity, is a subject which, though still involved in much uncertainty and doubt, requires the careful attention of the physician. Clear ideas as to the principles which determine the flow of electricity under various conditions will not, it is true, supply the want of clear indications as to the strength and duration of the current to be used, but they will remove the misconceptions and errors so prevalent in electro-therapeutical writings, and prepare the way for more rational modes of action.

The usual method of prescribing a current of so many cells cannot, unfortunately, be altogether eliminated from practice on account of the difficulty in obtaining measuring instruments; but every one who has obtained a fair understanding of the facts described and illustrated in the first chapter will feel how unsatisfactory it is.

Even the expressions 'very weak, weak, moderate, strong, and very strong, (see pp. 15, 31, etc.) used to designate currents of known absolute strength, though they eliminate the enormous fallacies arising from leaving out of sight the variable electromotive forces and resistances, must be taken in a very general sense indeed, and with reference to the average current-strengths used in medicine. They leave out of sight two important conditions which regulate the employment of electricity in individual cases, viz., the parts to which the current is applied, and the areas of the body-surface which form its point of entrance and exit. Thus, for instance, a "weak" current will usually be found too energetic when applied to the eye; and will cause much more pain when applied by means of a fine electrode than does a "very strong" one applied through electrodes of very large size.

Hence it must be understood that the expressions strong, weak, and the like, when applied to medical currents have no reference to their physiological effects or to the therapeutical aspect of the question. It is necessary that the spots of application and size of both electrodes should be indicated before any statement concerning the strength of current can have any meaning, beyond their effect on the galvanometric needle, as defined above. Throughout this chapter they are used in this limited sense only.

There is yet another consideration which makes an accurate electrical posology a matter of great difficulty: in electrisation it is usually only that part of the current which reaches the diseased structures which can be considered as having a curative influence. Now from what we know of the behaviour of currents in complex conductors—of their diffusion in the human body, for instance—it is obvious that this factor must introduce much uncertainty in all data on which we might base a generalisation. The only means of acquiring some precision in dealing with the subject is to combine a clear notion of absolute current-strength with rational methods of applying the electrodes. The latter point will presently be touched upon more in detail.

Finally, the universal habit, which has hitherto prevailed among writers, to express electrical doses in terms of cells of unknown strength makes all past experience of but small avail for the framing of a posology.

The quantity of electricity sent through the body depends upon two factors: the strength of the current used, and the time during which it is allowed to flow. One unit of electrical quantity (ampere) will have been administered, whether we have used a current of 1, 10 or 50 milliamperes during 1000, 100, or 20 seconds respectively. Is it indifferent to use a weak current for a long time or a strong current for a short time? Certainly if the chemical (electrolytical) effects of the current on the tissues alone are to be taken into account from a therapeutical point of view. Not so if its physiological effects come into play; for we know that not only there are limits to the excitability of the irritable tissues but also that the results of strong and weak excitations are frequently different, or even opposite. Now the therapeutical value of the current depends at least as much upon its influence upon the vital process of the tissues as upon their chemical constitution. Therefore the strength and duration of electrical application cannot be a matter of indifference. At the same time the ignorance in which we are still, as to the very complex conditions which determine the curative results obtainable from electrisation compel us to forego any *à priori* views on this subject and rely solely upon the data of clinical experience.

Now it is the unanimous opinion of all authorities that in the immense majority of cases everything electricity is likely to do is to be obtained by the application of moderate currents for a moderate time and repeated at intervals. In certain cases, however, "continuous electrisation," that is to say the application for long periods of very weak currents, appears to be beneficial.

We shall presently describe the methods of electrisation in detail, and discuss their rationale and the indications for their use.

CHOICE AND POSITION OF ELECTRODES.

The most ridiculous blunders are being constantly perpetrated by men learned in physic, but ignorant of physics, in their application of currents to patients. The old notion that electricity is a fluid to be poured into the organism is, if not openly professed, at least still too commonly acted upon. The fact that the current has density enough to exert any action near the electrode only, is too often ignored. The old mode of torture by putting a copper cylinder in each hand and sending tremendous shocks through the arms is, to my knowledge, still too often inflicted by doctors upon unfortunate patients for the relief of neuralgia, paraplegia, and many other such disorders. I have heard lately of a case of myelitis in which the application to the soles of the feet of the electrodes of a weak galvanic battery, was stated to have effected a cure!

The question is sometimes asked: "Is it not indifferent to apply the current anywhere with a large or a small electrode provided that the quantity of electricity be the same? Why should not this quantity be administered in the form of a galvanic bath instead of through a plate or disk?" The action of a given current is proportional to its density; there is no more resemblance between the effects of it when distributed over the whole surface of the body, and when applied to a limited area, than between the effects of a certain amount of heat diluted in the water of a bath or concentrated upon a definite region by means of a poultice. The size and position of the electrodes, that is to say, of the points of entrance and exit of the current, are therefore essential factors in the therapeutical dosage of electricity—which is not a fluid to be poured into the organism like so many ounces of a mixture, but rather an influence to be brought to bear with a definite intensity upon a definite portion of the organism.

Our object in applying therapeutically the current to the body being, we repeat, to act in a definite manner upon certain diseased structures, the main physical conditions to be fulfilled are, that the current be of such a strength, the electrodes of such a size and shape, and in such a position on the surface, that those structures be permeated by a current of the desired density. It behoves us then to be perfectly clear as to the principles stated in the first part of this book, (pp. 40 to 48) and which should guide us in carrying out these indications.

We subjoin here a number of rules designed to illustrate those principles in their application to practice:—

1. When it is desired to concentrate the effect of one pole upon a small superficial structure, such as a nerve or motor point (*e.g.*, in diagnosis), the electrode applied over it must be small. The other electrode, of large size, is placed upon a distant part of the body, (fig. 79).

2. When a deeper organ, such as the bladder, or a portion of the spinal cord, or of the brain, has to be reached from the surface, a large electrode, of appropriate shape (elongated in the case of cord), is to be applied to the surface over the organ. The other

FIG. 84.

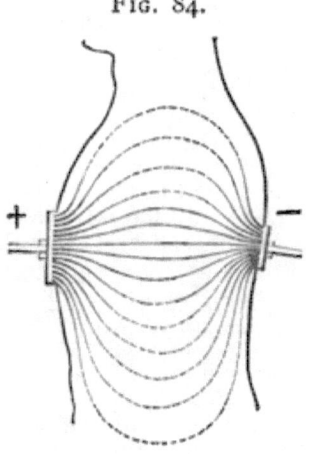

In this and the following figures, the areas of greatest current density are represented by full lines. The dotted lines show where the diffusion makes the current physiologically inactive.—In fig. 84 a very large electrode is placed on the sternum, and a large one over the vertebræ. The density is proportionally greater under the latter, (cf. fig. 30). Their relative position, opposite one another, ensures that the spinal cord be in the area of greatest current density.

electrode, of large size, is placed on the opposite side of the body, so that the organ be in the straight line joining the two electrodes, (figs. 84, 88).

3. An elongated organ of relatively large diameter, such as the cord (1) may be included between two large electrodes, by placing one over each extremity (bipolar method), and using a strong current so as to ensure the circulation through it of derived currents of a certain density (fig. 86); or (2) may be brought successively and by segments under the influence of one pole (unipolar method): one electrode, of large size, is gradually drawn and al-

FIG. 85.   FIG. 86.

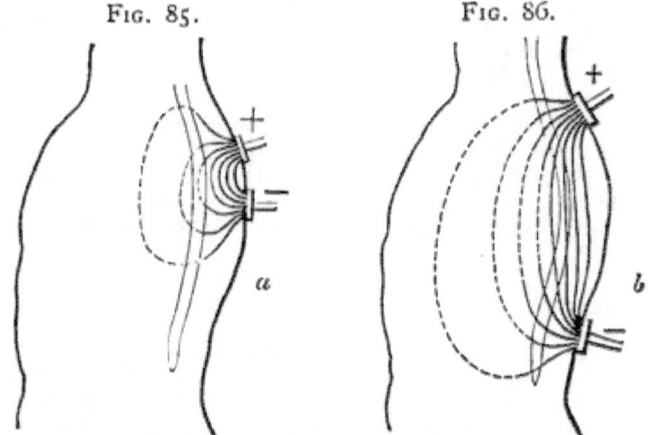

These figures show the necessity of placing the two electrodes (on the bipolar method) sufficiently apart, to allow derived currents to reach the organ to be influenced when the latter lies at a certain depth below the surface.

lowed to rest at several points along its whole length whilst the other is resting over one extremity of it, or better still on the opposite side of the body, (fig. 84).

4. When an articulation or segment of a limb, etc., has to be submitted to the influence of the current, two electrodes, each of sufficient size to cover an aspect of the part, are to be placed upon it opposite one another so as to secure its complete permeation, (compare fig. 88).

5. The electrisation of muscular masses is to be effected by placing upon them two electrodes of appropriate size, very large for the trunk muscles, large for the femoral or lumbar, medium for

FIG. 87.   FIG. 88.

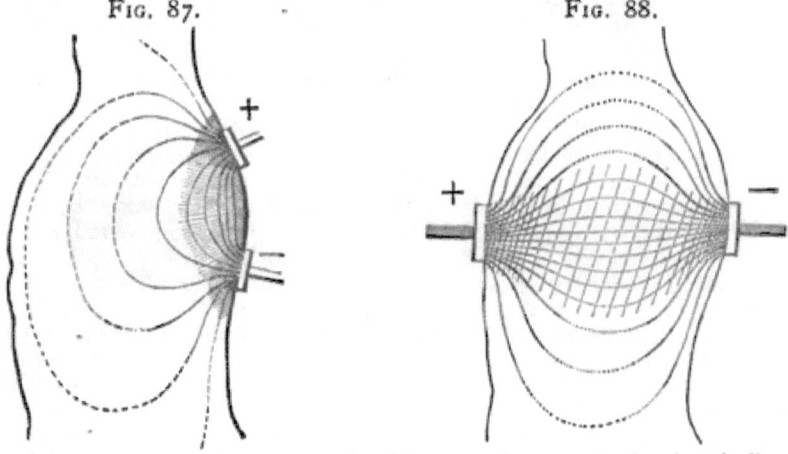

These two figures show by cross lines the areas of current density (i.e. of efficacity), when the electrodes are placed on the same aspect of the body and on opposite sides of it.

the arm and leg; or by fixing one large electrode to a convenient part of the body and successively bringing the whole surface of the muscles under the action of medium-sized electrode ("labile" method).

6. When very deep organs, such as the pelvic and abdominal viscera are to be influenced, two very large electrodes (with correspondingly powerful currents) must be applied so as to ensure that the particular part affected be in the area of greatest current density; viz., in the straight line joining the two electrodes.

PRACTICAL REMARKS ON ELECTRISATION.

Our electrical posology is subject, we repeat, to the same restrictions as the dosage of internal remedies; the idiosyncrasies of patients often make it imperative to depart widely from the quantities empirically laid down as indicated for the treatment of an average case of their complaint. Personal experience and medical tact alone can guide the physician's judgment in such circumstances. As a general rule, very strong and painful currents are not desirable. They often do more harm than good. At the beginning of a treatment, especially, great caution must be exercised in order not to "over galvanise." In our anxiety to cure quickly we often are tempted to err in this direction. Whilst some patients recoil at the very sight of the electrodes, not a few are under the impression that the more severe the ordeal, the more good it will do them and urge us to act in consequence. The silent action of mild galvanic currents is often rapidly successful where violent shocks and tetanising faradisation would prove disastrous.

With reference to the frequency of electrical applications, it is generally considered that one daily is usually required; in very chronic cases one, two, or three times a week may prove sufficient. It is only in certain cases of neuralgia and the like that they need be more frequent. The best rule to follow is to repeat the applications at sufficiently near intervals to obtain a cumulative action, so to speak.

We have two chief indications to direct us in the selection of the parts to which the electrodes are to be applied, viz:—the localisation of the disease, and that of the symptom. Thus in paralysis from cerebral or spinal lesions, it is found that both the nerve centre itself, and the paralysed nerves and muscles, may be beneficially influenced; and let us mention here, that for central applications the galvanic current alone is admissible, since our object is not to excite nerve action, but to modify nutrition by the so-called "catalytical" effects of the current. When applied to the peripheral organs, in the case of a central lesion, electricity can only act by a centripetal or reflex influence upon the centres, unless there be peripheral trophic changes, as well as paralysis, when local applications may reasonably be supposed to assist the natural processes of regeneration. Similar considerations hold when the symptom to be relieved is pain or spasm. If we know the actual seat of the primary disturbance, we must by electricity (or otherwise) attempt directly to correct the latter; but the symptomatic treatment should never be neglected, for it often proves when used alone of the highest benefit, even in cases where the symptom is of undoubted central origin.

In certain constitutional diseases, or when there exists a general vital depression of nervous or other origin, it is often useful to bring under the electrical influence the whole, or the greater part, of the body. The methods of "general electrisation," applied of late with so much success as an adjunct in the Weir-Mitchell plan of treatment of hysteria, are to be considered as tonifying measures, calculated to stimulate the functions and nutrition of the neuro-muscular system; the galvanic current being used for influencing the cerebro-spinal axis, the faradic current for stimulating the skin, nerves, and muscles.*

It should not be forgotten that it is not so much "electricity" which cures as "electrisation," that is, the rational and skilful application of electricity. The often-used expression "trying electricity" conveys about as much meaning as might that of "trying water"—without specifying whether the water is to be hot or cold, in the shape of ice or steam, applied externally or internally, and so forth. As the soothing influence of a hot bath, for instance, differs from the exciting action of a cold douche, so does the sedative influence of a mild, continuous galvanic current differ from the stimulating effect of an interrupted or faradic current. Where the one is beneficial, the other may be useless or hurtful.

It is certain that in many cases the patient injudiciously treated, has, besides suffering much unnecessary pain, found his symptoms aggravated; the fault lay then, not so much with electricity, as with the electriser. Every one who prescribes or applies electricity must pay the same attention to the administration of this powerful agent, both as to quantity and mode of application, as he would to that of any other therapeutical means.

It may be put down as a general rule, that whoever wishes to apply electricity rightly to others, should have first tried it upon himself. This practice of experimentation upon oneself, teaches far more quickly and surely than anything else, both what to avoid, and what to do in order to be a successful operator. It is indeed difficult to understand how galvanism and faradism can be intelligently, painlessly, and even sometimes safely, applied by one who has no personal experience of the effects and sensations produced by the currents. A general idea of the art of electrisation is very speedily obtained by a few personal trials and experiences. The different sensibilities of different parts of the body, the motor points of muscles, the mode of applying the rheophores to the skin,

---

* Much attention has lately been bestowed upon the value of dry faradisation (with the metallic brush) in certain lesions of the nerve centres. Should the results already published be confirmed by ulterior researches the practice of electrotherapeutics would be enriched with a valuable acquisition; whilst its theory would receive important confirmation with reference to the part played by reflex action in peripheral electrisation.

of graduating the current, etc., are by these means vividly realised and readily mastered. The numerous details in the application of electricity conducive to the comfort of the patient and efficiency of the treatment, can be learnt by experience only, and above all by personal experience. The sedulous practice of the experiments described in chapter II, will be found very useful owing to the familiarity with the manipulation of electrotherapeutical instruments which will be thereby acquired.

Avoid all unnecessary pain. The seat of pain is chiefly in the cutaneous nerves, and it will be found that where a strong current must be used, the application will be much less painful if the thoroughly moistened electrodes are held firmly pressed upon the skin; the more lightly they touch the surface, the more intolerable the sensory excitation. In nervous or thin skinned patients, in children, and whenever the head and neck are to be electrised, it will be found advisable to place the rheophores *in situ* before the current is turned on; and then to bring it very gradually up to the desired strength. Likewise at the end of the sitting it is best not to remove the electrodes suddenly, but gradually to diminish the current so as to avoid an unpleasant shock. It is a good policy with children and nervous people whom (at first especially) the very contact of the electrodes agitates, to wait a short time before turning on any current at all. Remember that neither the number of cells, nor the sensations of the patients are reliable guides as to the strength of the current flowing at any time. The galvanometer alone gives absolutely certain and definite indications.

Habit not only makes patients more confident, but enables them to bear stronger currents comfortably. Electricity in this respect is like many other therapeutical agents. In fact it is often necessary to increase the doses as tolerance becomes established. The skin, however, never gets accustomed to the chemical action of the current; and though the patient may not complain, unpleasant eschars are sure to follow too strong or too prolonged applications—especially if imperfectly covered or too small electrodes are used.

Whenever the electrodes are to be applied with the current already turned on, it should be made an invariable rule by the beginner to try it on his own hand, or face, as the case may be, before touching the patient, so as to avoid the danger of sudden shocks mentioned previously, (p. 83).

GALVANISATION.

There are three methods of applying the galvanic current, according as we wish to obtain its sedative, alterative or catalytical influence only, or combine the latter with its excitatory effects.

With the former object in view the current is made to flow continuously and steadily through the part affected. This method of applying it has been distinguished by Remak as *stabile*. In order to avoid shocks at the beginning and at the end of the application, the current must be gradually made and broken. For this purpose the best plan is first to apply the electrodes to the body, then by means of a proper collector (dial or sledge) to bring up the current to the desired strength. The collector ought to take not more than two cells at a time, and the switch moved gradually up to the full number of cells required. At the end of the sitting, likewise, the current is gradually taken off by returning the switch to zero. During the whole of the application the electrodes are to be kept firmly applied. These precautions are especially needed when the head and neck have to be galvanised. Nervous patients generally require this careful handling; for sudden jerks and interruptions are apt to upset them and make them dread the applications. A vivid shock produced by abruptly removing the electrodes tends also to destroy the sedative effect which continuous galvanisation is usually intended to secure.

When the battery has no proper means of graduation, and is fitted for instance with a simple pin and hole collector, stabile galvanisation may be secured by the following manipulation. Fix the rheophores so as to include the number of cells required; apply one of the electrodes *in situ*. Then with the other electrode complete the circuit by gradually applying it to some thick skinned or hairy part near the spot where it is intended to rest, and draw it slowly along until it rests upon that spot. For instance, if the face has to be galvanised, fix one pole to the back of the neck and apply the other over the hair above the temple and thence move it with gentle but steady pressure on to the skin, over the seat of the mischief. Repeat the process in the inverse order when the current has to be removed.

We have said that the current is sometimes used continuously for many hours, (p. 153). In former days this was the most popular method of electrisation, and used to be carried out by means of galvanic chains, bands, or belts, the survival of which at the present time is testified by advertisements in the daily papers, and the occurrence of eschars on the bodies of the believers in their virtues.

Continuous galvanisation is best effected by applying to the desired spots two well padded metallic plates, (p. 90), connected with two or three portable cells (small Leclanché or chloride of silver).

The *labile* method forms a transition between the stabile just described and the next. It consists in keeping one pole applied to one point, whilst the other is used for "sponging" the diseased limb, or portion of the body. This is effected by passing the electrode all over the surface to be acted on, with a moderate amount of pressure. Without necessarily producing any marked muscular contractions, labile applications are much more stimulating than the stabile. The circuit is not actually broken at any moment, but the current being densest in the part immediately beneath the electrode acts with an always changing energy upon the underlying muscles. The amount of muscular contraction produced by labile applications depends as much upon the rapidity of the motion, as upon the strength of the current used. It is obvious that the shifting polar zone under the electrode must necessarily act upon tissues which were just before under the influence of the peripolar influence of the opposite electrical sign (figs. 78, 79), and hence must be in a state of increased excitability as mentioned above, owing to the after effects of that polarisation. The "labile" method therefore owes part of its exciting influence on nerve and muscle to the effects of the mild voltaic alternatives to which it gives rise in the tissues. Labile currents are indicated where, in addition to the chemical action of the stabile current upon the deeper structures, we wish to call forth moderate contractions of all the muscles, and increase the cutaneous circulation over a wide area, and generally stimulate the nutrition of a limb.

There are two ways of using the *interrupted* galvanic current.

One of the poles is held immovable on one point of the body, whilst the other is dabbed over, or made to touch repeatedly, the skin of the part to be stimulated.

Or the two poles are both kept in contact with the skin, whilst the interruptions are made in the metallic part of the circuit by means of one of the contrivances we have described. When strong currents are to be used, the latter method must be preferred, being much less painful. It is the only reliable one for obtaining accurate results in electro-diagnosis. The interrupted current is stimulating, and is chiefly used for causing contractions of the muscles. The negative pole is to be applied over the muscle to be stimulated, being much more active than the positive, at least when the muscle is healthy.

Interruptions are never to be used about the head and neck without caution; giddiness, nausea, flashes of light are sometimes very easily produced, to the great discomfort of the patient.

When, the two poles being held in contact with the patient's body, the direction of the current is suddenly *reversed* by means of the commutator, the effects produced are much more powerful than those obtained by mere interruptions of the current, (see p. 107).

"Voltaic alternatives," as such current reversals are called, are indicated where a still more stimulating application than interruptions is required, and a corresponding degree of caution must be exercised in their use.

FARADISATION.

The subject of faradisation falls under two heads, according as to whether it is the skin or the deeper seated tissues that are to be acted upon.

Faradisation of the skin.—When we wish to localise the effect in the skin, it is necessary first to dry it thoroughly. This is best done by means of some toilet powder. The electrode used must be metallic, either in form of a solid cylinder, or of a wire brush, connected to the negative pole of the secondary coil (fine wire). The method is as follows: connect the positive pole to an ordinary moist electrode placed upon a distant part of the body, in the hand for instance, or on the back, or sternum; then lightly pass the metallic electrode over the dried skin, graduating the current so as to produce a lively crepitation, due to the passage of sparks between the skin and electrode. The current, however, must not be strong enough to produce any contractions of the underlying muscles. When not used for anæsthesia this mode of procedure is necessarily somewhat painful. When sensitive patients, or sensitive parts of the body, such as the face are to be so treated, it is advisable to use the hand instead of a metallic rheophore. The operator then takes the negative pole (moist) in his left hand, and with the back of his right hand, dried and powdered, strokes gently the patient's skin. The same crepitation is heard; and if the current is properly graduated, the sensation to both parties is rather pleasant than otherwise.

Faradisation of muscles and nerves.—When faradism has to be applied directly to the muscles through the skin (with well moistened electrodes), the easiest way is to hold the two handles in the right hand, the one between thumb and index, the other between middle and ring fingers, the upper end of the handles resting in the palm. The left hand is used for the graduation of the current, a most important point to be attended to, since different muscles require different current-strengths to react fully, and different parts of the body vary greatly in sensitiveness. A current well borne on the back of the forearm, for instance, will cause pain if applied to the front; bony prominences are very sensitive. The two electrodes are then passed up and down over the whole surface of the muscle, so as to ensure the thorough contraction of all its constituent bundles. We have already observed that in the case of many muscles, there is one or more points where the effect is much more marked, and that when one of the rheophores is upon it, the muscle is thrown at once into a general and powerful contraction; such points—the "motor points"—of the muscles correspond to the points of entrance of the nerves. The knowledge of these points, therefore, enables us to obtain the maximum effect with the minimum current strength and with the smallest amount of pain to the patient. When a whole limb or group of muscles has to be faradised we may conveniently use the "labile" method, as described under galvanisation; one pole is connected with a large electrode held immoveable over some distant part of the body, whilst the other is attached to a mounted carbon disk, with which the surface of the part is sponged or stroked with slow movement and firm pressure. The size of the electrodes must be small for the localised faradisation of single muscles, motor points, and nerve trunks; large for the excitation of flat muscles or groups of muscles. Thick wire coils are necessary for exciting large muscular masses, the abdominal organs, etc., for the reason indicated at page 75.

Faradisation of nerve trunk is resorted to mainly for inducing contractions in groups of muscles; or of muscles not to be excited directly. We thus faradise the phrenic and posterior thoracic nerves in the neck for stimulating the action of the diaphragm and serratus magnus. The method to be used here is that described as the unipolar (p. 147), one electrode being placed at a distant part of the body, the other on the motor point.

Faradisation of special organs will be found described in the special part under the different diseases where it is applicable. The various forms of electrodes necessary for this purpose are shown in the chapter on apparatus, (p. 92).

GENERAL ELECTRISATION.

UNDER this head we include central galvanisation, general faradisation, and electric baths.

Central galvanisation consists in bringing the whole cerebrospinal axis under the influence of the current. The negative pole being held stationary over the pit of the stomach or the sacrum, the positive pole is held first upon the forehead and vertex, then along the course of the sympathetic and pneumogastric in the neck, lastly, upon the cervical, dorsal, and lumbar regions of the cord successively. Here, as usual, a very careful graduation of current strength is needed during the application to the head and neck. Strong currents may usually be applied to the spine without fear. The duration may extend to 2 minutes to the head, 4 to the neck, and 10 to the spine. Central galvanisation is indicated in cases of general nervous depression, of nervous insomnia, etc., and is useful as an adjunct to local modes of treatment.

General faradisation consists in sponging the whole, or the greater part of the body with one electrode, whilst the other is held *in situ* on some insensitive part of the body, sternum, sacrum, etc., or fixed to a copper plate upon which the soles of the feet rest. The difficulty of the application lies in the proper regulation of the current necessary to stimulate fully, without causing actual pain, the different parts over which the sponge is being carried; this is best learnt by experiment upon oneself.

General faradisation by its action upon the muscular system promotes tissue change and nutrition. It braces up the system, and promotes circulation, as evidenced by a rise of temperature sensible to the thermometer. It is indicated in constitutional states where the vital processes are depressed, in hysteria, anæmia, dyspepsia, etc., and whenever the artificial muscular exercise it gives is likely to assist other modes of treatment. Where there is " irritable weakness" it should be used with caution, on account of the over excitation it is liable to produce. The application is to last about 10-15 minutes, or longer, special attention being paid to the thorough stimulation of the muscles of the neck and back.

The electric *bath* has been extolled for its manifold virtues. Much has yet to be done before its efficacy can be satisfactorily established; one point is certain, however, and that is the uselessness of the bath for local diseases. Both galvanic and faradic baths consist in immersing the patient in an ordinary warm bath through the water of which the current is sent. One question obviously arises: how much electricity penetrates the skin and actually passes through the body? This has never been answered by actual experiment. We shall only remark here that the usual method of putting salt or acid in the bath is a blunder, since the greater part of the current is thus derived through the water which thereby becomes a better conductor than the body.

An easy method of practising general faradisation by the patient himself, is to put the positive wire in the water of an ordinary hip or other bath in which he sits, and connect his sponge with the negative. If the coil is not strong, salt water must be used. The sponge is then carried over the whole surface of body; in order to prevent the current from flowing through the hand holding the sponge the latter should be held with an insulating glove made of some india-rubber sheeting or tissue. This simple application will be found a very useful substitute for the faradic bath.

GALVANO-FARADISATION.

UNDER this name I have introduced into electrotherapeutics a method of treatment which I have been in the habit of using for the last few years. It consists of uniting the secondary induction coil and the galvanic battery in one circuit, by connecting with a wire the negative pole of the one with the positive of the other, attaching the electrodes to the two extreme poles, and sending both currents together through the body. The current alternator and combiner described above enables the operator to unite the two currents with the greatest ease (p. 95).

The effects of the faradic current are greatly enhanced by a simultaneous galvanisation, because the points upon which the stimulus falls are in a state of exalted excitability or katelectrotonus (see page 111). Owing to the "refreshing" properties of the galvanic current upon muscle, the fatigue and exhaustion which might otherwise be the consequence of energetic faradisation are avoided.

Galvano-faradisation also effects an obvious saving of time in the numerous cases where both currents are indicated.

The future will show whether the simultaneous use of the currents has any advantage which cannot be obtained by their separate application. For the present we have not sufficient experience to decide this question; but were it upon grounds of convenience only galvano-faradisation must be warmly commended, when for instance strong contractions are desirable, or contractions of large and deep muscular masses. Electrisation of the abdominal organs is best carried out by its means. Wherever faradisation alone is likely to do good, galvano-faradisation is likely to do at least as much if not more. When a weak induction current only is at our disposition, the simultaneous use of galvanisation makes up for the deficiency. My own experience of the method in various rheumatic conditions, in atrophic paralysis, etc., has hitherto been very favourable.

With reference to the strength of each of the component currents in galvano-faradisation, it may be said that, as a rule, that number of cells which would be used for each case, if galvanism alone were applied, is to be employed. The faradic current is graduated according to the amount of muscular contraction or sensory stimulation desirable.

SUBAURAL GALVANISATION.

UNDER the title of "galvanisation of the sympathetic," there has been and is being accumulated a large amount of that kind of literature which has done so much to cast discredit upon the subject of electrotherapeutics.[*] Erroneous physics, imaginary physiology, fantastic pathology, are there worked by processes unknown to ordinary logic into a sort of mystical creed. One electrode is applied below the ear, the other on the nape of the neck, and a weak current passed. Though no physiological effects, no pupillary nor vasomotor changes are observed, ascribable to an excitation of the superior cervical sympathetic ganglion—which is under one of the electrodes—the curative results, real or imaginary, stated to have been obtained by this procedure in numberless diseases, have been ascribed to the electrisation of that ganglion. It is notorious that the sympathetic, like all other structures whose functions are still surrounded with a certain mystery, has ever offered an ample field for the growth of pseudo scientific imaginations. The reader of electrotherapeutical literature knows the deplorable energy with which this culture has been prosecuted in this particular department. Still, as in the case of other delusions, there is a certain substratum of fact underlying the absurdities perpetrated by the galvanisers of the sympathetic. A sober inquiry into the cases reported and unprejudiced clinical experimentation have, I think, shown that the position of the electrodes in subaural galvanisation is favourable to the therapeutical results in many cases. It would not, however, be rational to imitate the example of those we have just condemned, and offer any theory to explain the results obtained. We may safely say that it is not the sympathetic which is the sole, or even the chief, channel through which the current acts beneficially, but we cannot on the other hand define the part played by any one of the numerous nervous structures reached at the same time by the current (base of brain, spinal cord, pneumogastric, cutaneous nerves, etc). We adopt subaural galvanisation as an empirical mode of treatment, as we do for instance the application of various substances to the skin for the relief of symptoms localised in deep seated organs : to account for the effect of the latter by an appeal to reflex actions is to throw but a thin veil over our complete ignorance concerning the essential phenomena of the process.

The methods of subaural galvanisation will be mentioned again in the paragraphs on the treatment of the brain and spinal cord. The usual method is to apply the kathode (medium electrode) under the ear, and a large flexible plate over the lower cervical and upper dorsal vertebræ. Weak current (6—14 cells) for three or four minutes on each side.

[*] "An Electro-therapeutical Superstition," by A. de Watteville—Brain, vol. iv., 1881, p. 207. The term *subaural galvanisation* was there proposed as a substitute for the offensive expression "Galvanisation of the Sympathetic." I retain it here the more readily that Erb (Elektrotherapie, 1882, p. 258), recognises its appropriateness.

B. SPECIAL ELECTROTHERAPEUTICS.

The concluding part of this chapter is intended to illustrate the application of the principles exposed in the former portions of the book to the treatment by electricity of various morbid conditions. The pathology and clinical characteristics of the diseases which lend themselves to electrotherapeutical procedures are scarcely touched upon here, the reader being supposed to have mastered one of the usual text-books of general medicine. The curative value of electricity differs greatly, even among cases of the same nature, according to the amount of the alteration present in the tissues submitted to its influence; its use must be guided by the same general considerations, and is subject to the same prognostic conditions, as that of other agents.

HYPOKINESIS:—PARALYSIS.—PARESIS.

The commonest symptom for the relief of which electricity is prescribed is paralysis, or paresis as it is conveniently named when incomplete. This symptom is the consequence of numerous morbid states, organic or functional, of the brain, cord or nerves. It is to be expected *à priori*, therefore, that the results of electrical treatment will greatly differ according to the seat and nature of the lesion which gives rise to paralysis.

The disturbance may be functional or molecular, that is, involve no visible structural change in the tissues; or proceed from the presence of some poison in the system; or be of an inflammatory or degenerative nature. Its seat may be the nerve cells and fibres themselves, or the connective tissue which supports them. The blood-vessels or membranes may be primarily at fault. Solutions of continuity in, or mechanical pressure upon, the nervous elements may be the cause of the paralysis. It is obvious therefore that in many cases, including all those where the destruction of nerve-tissue is complete and irreparable, electrisation will fail to do any good. In others it will come into service only after the immediate causal indication has been fulfilled. In the rest, finally, it will form the chief, if not the only means of treatment to be adopted. The following remarks are intended to give a general idea of the rationale of the so-called anti-paralytic effects of electricity, and of the methods of applying it for the relief of the symptom and its cause.

Electricity acts in two ways: directly upon the nutritive processes and irritability of the tissues which are permeated by the current; indirectly by centripetal or reflex action upon these tissues and upon the centres which preside over their innervation. Its direct action is attributed to the "catalytical" and "modifying" influences above described (p. 147); its indirect action to the exciting property of the current upon nervous structures. Hence an electrical treatment must consist of applications both *in loco morbi* and *in loco symptomatis*.

A. The application of electricity to the seat of lesion in cases of paralysis of central origin is perhaps too commonly neglected. It is true that we have too often but a very imperfect idea of those morbid processes in the nerve centres upon which the symptom depends, and that we have no right to assume that the current has any specific curative influence upon any one of them; still as a justification for central treatment in such cases we may plead our very ignorance, we may urge the poverty of our therapeutic arsenal in arms wherewith to combat the enemy; and may also invoke the possibility of at least staying its progress by promoting nutrition of the surrounding portions of the nervous structures threatened by its invasion. Where a destructive lesion has occurred we may not unlikely restore the functions of some of the involved nerve structures which have survived, but the activity of which remain in abeyance, or stimulate the functions of other neighbouring structures suspended in some instances by sympathy as it were, by shock or inhibition. Finally, and chiefly, clinical experience seems to point to the reality of the results attributed to central electrisation in some cases of paralysis.

When the lesion is situated in the course of a peripheral nerve, the application of electricity *in loco morbi* is not only much more easy to carry out thoroughly, but its effects are perhaps more readily accounted for. It is then possible also to bring to bear another influence upon the morbid process, viz., that of artificial excitations applied to the nerve above the lesion. In this way it may sometimes be possible to overcome the resistance encountered at that point by the voluntary stimuli, and facilitate the transmission of the latter (Erb.)

FIG. 89.

n, Nerve; m, Muscle; a, Seat of Lesion. An excitation applied at e strengthens the effect of the voluntary stimulus represented by the arrow in overcoming the resistance at a.

B. We must carefully distinguish when attempting to give the rationale of the treatment of paralysis *in loco symptomatis* between the classes of cases which electrodiagnosis has enabled us to establish. In the first class the reactions of the nerves and muscles are normal, and remain normal as long as the primitive lesion (situated above the grey matter of the cord) remains uncomplicated. Here the nutrition and conductivity of the peripheral organs being intact it is obvious that any result which their electrisation can give must depend upon an influence extending upwards into the diseased centre. This influence, physiology teaches us, may be exerted through the channels of the sensory nerves; reflex actions and the reflex tonicity of muscles are familiar instances of it. It has also been proved that electrical excitations of a peripheral nerve, the sciatic nerve for instance, give rise to changes in the medulla oblongata. Finally, there is evidence to show that the motor nerves themselves, or perhaps their terminal organs more particularly, exert an influence on the nutrition of the cord. It is therefore admissible that the modifying influence of the current upon the peripheral organs innervated by a diseased centre, sets up certain changes in that centre, which may account for the results obtained. Reflex vasomotor effects in more or less distant and deep organs follow the excitations of the surface of the body—the whole theory of counter-irritation rests upon their existence; it is possible, therefore, that the artificial excitation of the peripheral motor organs and of their integuments acts upon their central organs of innervation by modifying the circulation in the latter.

When in addition to the paralysis there are disturbances in the nutrition of the peripheral organs it is obvious that electrisation of the latter may be of use by its direct effect upon them. Such disturbances may be so slight as scarcely to deserve more than the name of "functional" arrest of nerve conductivity, or may assume a more definite organic character, the varieties of which are numerous both in kind and degree, starting from simple hyperæmia and culminating into degenerative atrophy. Here the exciting, modifying and catalytic effects of the current may, if not explain, at any rate give a clue to its mode of action upon the altered and paralysed nerves and muscles. It is obvious however that unless the centrally placed lesion becomes *pari passu* the seat of a recuperative process, any local effect of the current will fail to bring about a restoration of the motor power.

When speaking of centripetal or reflex action I mentioned the case only where the mischief was actually central. But under certain other conditions, as for instance when the motor paths alone are the seat of a morbid process it is conceivable that excitation of the sensory termination of the reflex arc may also have some influence in the restoration of the motor conductivity. Finally it must not be overlooked that electrisation whenever applied to the surface, acts upon the circulation of the underlying tissues by exciting the cutaneous nerves. This factor naturally varies in importance with the method used, and is least when simple galvanisation is resorted to, much greater in importance when dry faradisation is applied.

From these various considerations, and in accordance with the teachings of the accumulated experience of many years' observation, the following general rules may be reduced.

1. In the electrical treatment of paralysis the suffering part must be submitted to the influence of galvanic currents, very weak and stabile in the case of the brain; stronger, stabile or labile, in the case of the cord or peripheral nerves. Both poles may be used successively.

2. When possible, excitation of the conducting paths above the seat of lesion by means of faradisation or interrupted kathodic galvanisation is to be applied.

3. Peripherally, labile kathodic galvanisation of the nerve trunks and muscular masses, or moderate faradisation, is to be carried out, when there is no trophic disturbance. Otherwise more energetic labile galvanisation, or galvanofaradisation is indicated; the kathode and anode being applied successively to the organs.

4. Cutaneous excitations with the faradic brush over the seat of the lesion or at a distance may be beneficial. They are more especially indicated when the motor symptoms are accompanied with sensory, or depend upon central vasomotor, disturbances.

HYPERKINESIS—SPASM.

By hyperkinesis is meant a state of the nerve cells or fibres which gives rise to involuntary muscular contractions. These spasms are divided into clonic and tonic according as they are of short or long duration. Clonic spasms merge into simple tremor on the one hand, and into convulsions on the other; whilst tonic spasms when permanent are also known under the name of "contracture."

Tremors are symptomatic of numerous conditions of the system at large, as well as of the nerve centres. They supervene in senile decay, degenerations of the cerebro-spinal axis, in asthenia, intoxication by various substances, paralysis agitans, &c.

Convulsions are often pathognomonic of the great neuroses, called functional, such as epilepsy, hysteria, etc., which will be mentioned further on. We shall here consider more particularly the localised forms of spasm, though the remarks concerning their etiology and pathology, and the therapeutical indications derived therefrom, apply equally to all.

The reader must be reminded that under the heading of contractures are included only muscular shortenings of neuropathic origin. Myopathic contractures (*i.e.* due to disease of the muscles themselves) and those arising secondarily from skeletal malformations form a very distinct group which might be appropriately distinguished as "muscular retractions." Contracture may be sometimes considered as an exaggeration of the normal reflex tonicity of the muscles.

The pathogeny of spasms is one of the most obscure problems of neurology. They are generally symptomatic of some neurosis, or functional disturbance of the nervous system; and to the fact of morbid anatomy being silent with reference to them must be attributed the uncertainty which still prevails concerning the localisation and nature of their causes. Generally the centres seem to be primarily affected; sometimes, as in tetany, we have to do with an alteration of the peripheral motor paths.

In some cases however (lateral sclerosis, secondary or idiopathic), our knowledge is more precise. Here the hyperkinesis depends upon a sclerotic process of certain motor paths, whereby the inhibitory influence of the higher centres being cut off the reflex activity of the lower ones is exaggerated; but which also acts probably as a direct excitant of muscular action.

Finally spasms appear to be caused in many instances, especially when affecting muscles connected with organic life, by undue reflex excitations of the motor centres. It is not easy however to prove that the result depends solely upon an excessive stimulus, and is not due to hyperexcitability of the nervous grey matter where the centripetal is converted into a centrifugal current. The starting point of the reflex excitation may be situated in some very distant organ or portion of the surface of the body.

The main general indications for the treatment of spasm will therefore be, 1. to find out and remove the cause; 2. to modify the hyperexcitability of motor nerves or nerve centres, or rather the functional or molecular disturbance upon which it depends; 3. to check the sclerotic or other morbid processes which may be present; finally, 4. where the primary cause of the reflex contraction, that is the peripheral stimulus which gives rise to it, cannot be discovered and remedied, to substitute for it another peripheral excitation more powerful than itself, or which will act upon the reflex centre as an inhibitory influence.

It is obvious, considering, the numerous and very different morbid states of which spasm is symptomatic, that no one agent or mode of treatment is applicable to all. Electricity now and then yields brilliant results, but also too often fails to cure or relieve even in cases where previous experience led us to expect success. Like all other "antispasmodics" its action is uncertain, and we often have resort to it more on account of the failure of every other medication than from any positive knowledge concerning its action in this respect.

The following methods must be tried successively, or in various combinations in the treatment of spasms:—

1. The nerve centres must be treated according to the methods indicated for paralysis, with the object of modifying their nutrition, whenever we have reason to believe that spasm depends upon an abnormal condition of the latter (see p. 174 *et seq.*).

2. Nerve trunks may be submitted to the action of stabile galvanic currents of moderate strength. The positive pole is to be tried first, then the negative, or both poles may be placed on the nerve.

3. Nerves and muscles may be subjected to "swelling" faradic currents: *i.e.*, currents slowly brought up to the maximum strength tolerated, and very gradually reduced again.

4. Voltaic alternatives may be made with strong currents in the affected nerves and muscles. My experience confirms that of other observers as to the value of this method, *e.g.*, in facial spasm.

5. Excitation of more or less distant parts of the body or of the integuments over the seat of spasm are useful in the reflex variety, as a corrective or inhibitory influence on the centres. The centripetal effects of the current described under paralysis may also be of use in the class of spasm due to central hyperexcitability.

6. Secondary indications are yielded chiefly by the presence of

points tender to pressure, or of points pressure upon which effects an arrest of the spasm. Weak stabile anodic galvanisation of, or cutaneous faradisation over, these points (which must be diligently sought for, even at a distance from the seat of the symptom) may prove useful.

The relaxed state of the antagonist muscles may furnish an indication to be fulfilled by moderate faradisation of them.

As a rule it is well here as elsewhere to try the gentler measures first, and failing these to proceed to the more energetic applications. The physician's personal experience alone will prove of any assistance in guiding him in the management of individual cases.

Trismus is usually of peripheral reflex origin. When due to a functional cerebral derangement galvanisation of the inferior portion of the motor area (see fig. 90) would be indicated.

Facial spasm (histrionic) is more commonly idiopathic. Simple galvanisation of the nerve and muscles may be tried, but will rarely be found of great use. Good results have been obtained from galvanisation of the centres. The anode (a large flexible plate) is applied two or three inches above the ear (see fig. 90) so as to cover the lower portion of the ascending frontal and parietal convolutions whilst the kathodic is fixed to the nape of the neck. A weak current (to be very gradually made and removed) is then allowed to past for five to ten minutes. The same application may be made also with the poles on the occipital region and the face respectively. Peripheral treatment may be alternated or combined with central, in the form of voltaic alternatives or swelling faradic currents through the nerves and muscles. Pressure points must be looked for over the area of distribution of the trigeminus, and along the cervical spine; when found they are to be dealt with as stated above.

In *Blepharospasm*, electrical treatment must be carried out on the same lines as when the whole facial nerve is involved.

Torticollis is a spasm of the neck muscles of peripheral origin and usually described as a rheumatic affection. It usually yields very rapidly to strong faradisation of the affected parts, or to the application to them of a galvanic current with interruptions or voltaic alternatives, at least in recent cases.

Spasmodic wry-neck is often localised in the muscles supplied by the spinal-accessory. Its treatment is too often palliative only. Galvanism should be applied (as in facial spasm) to the parietal and occipital regions of the head, and the various peripheral modes of electrisation above described brought successively into use if necessary. The prognosis is perhaps less gloomy where pressure points can be discovered; to them the anode of a current of moderate strength (15-20 cells) should be applied for five minutes daily.

Cramps and spasms occurring in the muscles of the arms and legs must be treated according to the general principles enunciated above. Among those affecting the *respiratory apparatus* may be mentioned hiccup, sneezing, and coughing. Erb recommends energetic cutaneous faradisation of the epigastrium; failing which galvanisation of the throat or phrenic nerves may be tried. In an old standing case of excessive sneezing I was greatly astonished to see lasting and complete relief given by two applications of the galvanic current to the mucous membrane of the nose with the electrode shown in fig 70.

ANÆSTHESIA.

ANÆSTHESIA depends upon a functional or organic disturbance of the sensory organs of reception, central or peripheral, or of those of transmission between the terminal structures. It may affect any of the special senses, or the sense of touch (skin and mucous membranes), or the so-called "muscular sense", or any of the subsidiary forms of sensation (temperature, pressure, pain, visceral sensations, etc.)

The general principles governing the rational treatment of anæsthesia are the same as those stated under the head of paralysis. The causal indications are to be fulfilled as much as possible; treatment *in loco morbi* must have the precedence whenever the lesion can be localised, though in many forms of anæsthesia the mere symptomatic indications and applications *in loco symptomatis* are empirically found to be successful. Electrisation is more commonly directed against the usual forms of loss of tactile sensation, from peripheral impressions (cold, injury, etc.), or from certain cerebral disorders (hysteria, congestion, hæmorrhage, etc.). According to the seat and nature of the pathological process which gives rise to the symptom, one or more of the following general indications will have to be fulfilled. Firstly, treatment *in loco morbi* to remove the cause. This is to be carried out on the lines laid down in the paragraph on paralysis:—Galvanisation of the brain, cord, nerve, or skin, so as to improve the nutrition of the part by the catalytical influence of the current, or to increase the depressed excitability of the structures. Secondly, treatment *in loco symptomatis* in order to supply artificial stimulus. Dry faradisation with the wire brush fulfils this indication admirably. It may be that such excitations act by overcoming obstacles in the sensory paths (cf. fig. 89) but their action on the local circulation, etc., may have a direct influence upon the nutrition of the ultimate nerve fibres and sensory organs.

The methods of galvanisation will be those already described of stabile applications to the brain and spinal cord; and of labile kathodic excitation with strong currents of the anæsthetic part and nerve trunks supplying it, the anode resting upon the back at the level of the points of origin of the nerves. Dry faradisation is to be applied in the usual way, *i.e.*, a moist electrode is fixed to some distant part of the body, and the well-dried part stroked lightly with the wire brush. The secondary coil (fine wire) is to be used, the current being strong enough to produce distinct crepitation or sparks. The local applications should be made every day for about ten minutes.

In hemianæsthesia (*i.e.*, loss of sensation on one side of the body), which is observed in certain cases of disease in one hemisphere, in hysteria, etc., Vulpian[*] has shown that it is not necessary to excite the whole anæsthetic surface, but that the energetic dry faradisation of a small area of the arm, for instance, may restore the function of the whole.

HYPERÆSTHESIA.—NEURALGIA.—PAIN.

CONDITIONS in which sensibility is exalted (hyperæsthesia) or perverted (paræsthesia) are rarely the object of a special electrical treatment. The presence of pain, (dysæsthesia, algesia) on the other hand, frequently indicates electrical applications which will naturally vary according to the cause and clinical characteristics of the symptom. Foremost stands neuralgia in its different localisations and manifestations. The purest types of it are the so-called idiopathic neuralgiæ, characterised by spontaneous paroxysmal pains in the area of distribution of a nerve trunk which displays no alteration appreciable to the eye; and symptomatic neuralgiæ, in which similar phenomena appear to be connected with one of the more general neuroses (hysteria, etc.), or with a diathesis or state of the general system, or with toxic influences. Many cases usually classified in the latter category must however be distinguished from the truly neuralgic class, and be spoken of as "neuralgiform" or "pseudoneuralgic," as would be the pains accompanying many forms of nerve-degeneration, sclerosis, or inflammation, as well as mechanical or traumatic lesions of the nerves.

The pains of spinal irritation, cephalalgia, myalgia, often yield symptomatic indications for electrical treatment; those connected with gouty and rheumatic conditions may likewise become its objects. In numerous morbid conditions (digestive, cardiac, pulmonary, or uterine disorders, etc.), there exist spots of sympathetic or referred pain; whilst painful pressure-points along the vertebral columns and the course of nerve trunks are the constant accompaniment of various organic or functional disturbances either of the nervous system or of other organs.

Electricity will often be found of the highest value as a palliative at least in the treatment of points of spontaneous pain; whilst tender pressure points have been shown by several observers to constitute important indications for local applications of more than symptomatic efficacy. (Cf. p. 169, etc.).

In some instances electricity may be advantageously used for the causal treatment of pain (certain neuroses, neuritis, etc.), and will then have to be applied according to the principles and rules given elsewhere in this chapter. In pure neuralgia, which we suppose to consist in a slight molecular or nutritive disturbance of the nerve

---

[*] This distinguished observer states that where aphasia co-existed with the anæsthesia there occurred a temporary improvement in the patient's speech after the cutaneous faradisation. This fact illustrates clearly the importance of the centripetal and reflex effects of peripheral electrisation.

fibres, this agent acts by directly removing the condition which gives rise to the symptom and so fulfils the causal indication. The general line of action to be adopted in the treatment of neuralgia is as follows.

Apply the positive electrode, as large as the circumstances allow, over the seat of pain. Use a weak or moderate galvanic current on the strictly stabile method (page 158) avoiding all makes and breaks and even any sudden change in the current strength. The negative pole is to be placed either over an indifferent point (sternum, etc.), or over the central origin of the nerve (cord). Duration of current, four to eight minutes.

If not rapidly successful, apply both poles successively to the painful spot; or place both poles on the nerve, positive above, negative below, then reversing the order and for the rest proceeding as first described. In obstinate cases try the effect of voltaic alternatives or of 5-8 minutes faradisation of the nerve with moist electrodes; and finally that of dry faradisation of the skin.

Tender pressure points must always be carefully sought for, and if any are discovered along the course of the nerve, or along the vertebræ, the anode (or the anode and kathode successively) is to be applied over each for three or four minutes, stabile and weak current.

When the pain is diminished by each application, but returns within a few hours, several short applications during the day may be safely resorted to.

The symptomatic treatment of the various other forms of pain admitting of electrical treatment must be carried out on the same lines as those just laid down for neuralgia, the strength of the currents being increased according to the seat and nature of the pain, and the susceptibility of the subject. The application of electricity for the relief of the fixed pains of dyspeptic, anæmic or phthisical patients, for instance, deserves an extensive trial though it has hitherto been too much neglected. The amount of suffering these symptoms involve, the uselessness of the usual treatments, and the rapid results often obtained from galvanisation, alike commend it to the attention of the general practitioner.

In the ordinary forms of neuralgia of the *Trigeminus* the results of electrisation are sometimes as rapid as gratifying; in true tic-douloureux, on the other hand, the opposite is true. The current must be localised chiefly upon the points of exit of the several branches of the nerve from the skull. The root and ganglion of the nerve are so deeply placed that it is impossible to act upon them with much effect, hence perhaps the inefficacy of electrisation in certain cases where the morbid modification of the nerve is perhaps localised not at its periphery but nearer the centres. In such cases currents as strong as bearable may be tried from forehead to occiput, through both temples, and both mastoid processes.

For ordinary cases a weak current (6-10 cells) applied for five minutes, several times a day if necessary, is to be used; the anode is placed on the points of emergence, the kathode behind the neck. The current must be gradually made and broken, according to the stabile method. If relief is not very quickly given, the other modes of electrisation previously described must be tried.

Cervico-occipital must be dealt with on the same principles as facial neuralgia.

In *cervico-brachial* neuralgia stronger currents (20-30 cells) may be used. The kathode rests on the cervical region and stabile applications of the anode are made along the painful nerve trunks. Labile galvanisation with the kathode of the whole arm is useful in cases where there are paretic symptoms. Tender pressure points along the vertebræ, cervical plexus and brachial nerves are to be diligently sought for and treated.

Intercostal neuralgiæ are often due to organic changes in the nerve or its surroundings and then resist every form of treatment. Such for instance is often the case in the pains of the herpes of old people. The treatment of intercostal neuralgia as a rule requires rather energetic measures such as alternate galvanisation and faradisation, or better still galvanofaradisation, one pole resting on the spinal origin of the nerve, and the other placed upon several points along its course. Moderate or strong currents (15-30 cells).

In *sciatica* the main point is to bring the whole of the suffering part of the nerve trunk and branches under the direct influence of the current. For this purpose; 1. a large plate is held over the sacrum or over the sciatic notch, whilst a middle sized disk is placed successively upon a number of points along the nerve and its painful branches; or, 2. two disk electrodes about six inches apart, are placed along the same, so as to bring its whole length successively and by segments, under the electrical influence. Twenty to forty cells are to be used according to the tolerance and fatness of the patient, for eight or ten minutes altogether. In addition to this a more or less energetic labile galvanisation or galvanofaradisation, of the whole limb is to performed whenever the patient is found to bear it, especially where there are signs of muscular atrophy and weakness.

Electropuncture of the nerve, and galvanisation of it through a rectal electrode have proved useful in certain cases, but should not be resorted to except by operators familiar with the electrolytical effects of the current. Simple acupuncture, however, presents no danger and may be safely tried in conjunction with percutaneous electrisation.

ELECTRISATION OF THE BRAIN.

AMONG the cerebral disturbances in which electricity may be used as a remedial agent some are purely functional, as chorea, hysteria, epilepsy, and other such neuroses, as well as many conditions vaguely described as neurasthenia, insomnia, cephalalgia, etc.; others again, are of a vasomotor nature (anæmia, hyperæmia). Hæmorrhage with more or less extensive pressure upon, or destruction of, brain tissue and its softening from arterial plugging, form another category of cerebral lesions within the reach of the electrotherapeutist. Finally he may have to do with sclerotic or degenerative changes of brain-cells and fibres. Cases of cerebral tumours and the like never present any indications for electrical treatment.

It is obvious that in the electrical treatment of brain disease we must expect to fail very frequently. But the gravity of disturbances affecting the source of all the higher life of man and the occasional striking successes obtained by this empirical mode of treatment, not to speak of the paucity of our therapeutical resources in such cases, makes it not unfrequently a duty to have recourse to electrisation before pronouncing a case hopeless. In cases where the loss of substance is manifestly considerable the possible improvement in the patient's condition is very small indeed; in others the benefit is more or less marked; in a few, too few unfortunately, it is as striking as rapid; in such cases the arrest of function must evidently be out of all proportion to any organic lesion that may be present. Indeed, all we can hope to do in many cases of organic lesion whether by electrisation or other methods of treatment is *to reduce the symptoms to their anatomical basis*.

The possible mode of action of the current on the nutrition and circulation of the brain, and on the excitability of its elements will be gathered from what has been said in the paragraph on paralysis.

The usefulness of cephalic electrisation in certain morbid conditions has been demonstrated by the results obtained in some well observed cases (cf. Erb's *Elektrotherapie*, lecture xvi). We therefore need not take any heed of the aprioristic objections urged against the practice by certain authors. The great principle which governs electrotherapeutics, that of treatment *in loco morbi*, is to be applied to its fullest extent in cerebral galvanisation, and the action of the current concentrated as much as possible either upon the cortical areas to which our present knowledge allows us to localise the lesion, or upon the deeper portions of the encephalon by a proper disposition of the electrodes.

For the purpose of electrising the brain the galvanic current should be used; the electrodes should be large, and made to fit closely the surfaces to which they are applied, the flexible plates described previously (p. 90) answer this purpose very well; and should be placed as to secure the utmost density in the region of the lesion, according to the rules laid down at the beginning of this chapter (p. 153). The current should be very weak, or weak (4-12 cells according to the state of the battery and the resistance of the patient's tissues); duration to be limited to one to three minutes at a time; cautiously increased if the subject bears it well. No sudden alterations in the strength should be made (stabile current). The following general directions for galvanic applications to the head will be found useful.

1. Longitudinal galvanisation; one large flexible plate on the forehead, another on the occipital region.

2. Transverse galvanisation; one large or medium plate on the mastoid region on each side.

3. Subaural galvanisation; a medium sized electrode under the ear, the other on the opposite side of the head, forehead, or nape of neck.

4. Localised galvanisation; one medium electrode over the actual convolution, etc., supposed to be the seat of the lesion, the other on the opposite side of the neck, (oblique galvanisation, Erb) or in the subaural position.

Two or more of these modes of application may often be advantageously combined. In most cases of cephalic electrisation the respective polarity of the electrodes seems to be a matter of indifference. There occur instances, however, where at least the subjective effects vary with the relative position of the poles; it will therefore be wise on the part of the physician to be guided here by the results, rather than proceed on a routine or a theoretical plan of action.

The electrical treatment of *cerebral hæmorrhage and softening—hemiplegia*—consists chiefly in galvanisation of the damaged portion of the brain, and galvanisation or faradisation of the paralysed muscles, according to the principles laid down previously. It should not be undertaken, in cases of ordinary severity, before a month has elapsed since the attack, and the tolerance of each patient carefully tested before currents of average strength used in his case. As a rule it will be found that unless improvement is apparent, after a week or two, no further good will be done by persevering in the treatment; and conversely that if progress ceases to become manifest after the first few days' electrisation, the case is one unsusceptible of further amelioration.

The supervention of "late rigidity" in cerebral hemiplegia, is an event of bad omen and makes the prognosis gloomy. It usually depends upon secondary descending degeneration of the pyramidal tracts—sclerosis of the lateral columns of the spinal cord. It is very rare that any marked relief is then afforded to the sufferer by galvanisation or faradisation, which should be applied according to the principles laid down in the paragraphs on spasms and spinal sclerosis.

When anæsthesia is present it forms an indication for an appropriate electrical treatment, which not rarely proves successful against the symptom (p. 171). Hemiplegic limbs are often the seat of pains and trophic alterations (œdema, glossy skin, arthropathy, etc.), which peripheral galvanisation may mitigate, though such accidents usually point to an irremediable destruction of brain matter.

Aphasia should be treated by localised galvanisation, one electrode resting over the region of the third frontal convolution, and island of Reil. (See Diagram of "Motor Points.")

The best method to proceed in an ordinary case of hemiplegia is, first, to apply a large electrode to the nape of the neck, and place the other, also of large size, on the diseased side and so as to include the seat of the lesion between the two, *i.e.* over the inferior fronto-parietal region. The hair must be well moistened, and the current regulated according to the rules given above. The polarity of the two electrodes is, according to the ordinary method, negative to the head, positive to the neck; when unpleasant sensations are produced by the current on the brain, the reverse disposition of the poles should be tried. Secondly, the kathode is applied in the subaural position, the anode remaining fixed to the nape of the neck; the current should be weak (6—14 cells), made and broken very gradually; duration three minutes on each side. Finally, the paralysed nerve trunks and muscles, should be submitted to a labile application of the galvanic current for 5—10 minutes according to the endurance of the patient. The current should be strong enough to be felt and produce slight muscular contraction, and the number of cells used regulated accordingly. The resistance of a hemiplegic arm and hand is often higher than the normal (especially when cold and œdematous), and comparatively higher battery power required. The kathode (medium sized disk electrode) is applied to the arm, whilst the anode remains, as before, fixed to the nape of the neck.

Simple faradisation of the paralysed extremities, face, tongue, etc., may be resorted to instead of galvanisation. In many hopeless cases where the patients notwithstanding insist upon having an electrical treatment carried out, they may be safely entrusted with a small induction coil and allowed to use it at home.

In a number of cases Prof. Vulpian has obtained good results from cutaneous faradisation. His method was to apply the wirebrush daily for ten minutes to a circumscribed area of the arm, using a rather strong current.

In *monoplegia* (and *monospasm*) of cortical origin the chief indication is to act upon the nutrition of the affected convolution, and for this purpose the galvanic current may in some cases be applied locally over the seat of the lesion by means of a middle sized electrode, (the other resting upon some other part of the body), which should be made successively anodic and kathodic, but without sudden interruptions of the current. The monoplegic limb may also be galvanised by the labile method as just described.

In the conditions known as "*cerebral anæmia*" and "*neurasthenia*," electricity has frequently been successfully employed. The following applications are indicated:—1. Very weak galvanisation of the head, one electrode resting on a distant part of the body, the other successively brought into contact with the frontal, sincipital, and parietal regions so as to act directly upon the whole cortical substance; or the electrodes being applied in the longitudinal, transverse, oblique, and subaural positions. All sudden interruptions to be avoided and the relative polarity of the electrodes to be determined empirically for individual cases.—2. General faradisation, including especially the frontal and occipital regions; and galvanofaradisation of the abdomen whenever there is visceral torpidity.

Against cerebral *hyperæmia* good results have been ascribed to, firstly, longitudinal galvanisation of the brain with the anode on the nape of the neck; secondly, cutaneous faradisation of the neck, trunk, and arms with moderately strong currents. Certain *head-*

aches and *sleeplessness* are amenable to galvanisation and faradisation according to the methods described under "neurasthenia."

In *bulbar paralysis*, that is to say disease of the motor nuclei of the medulla oblongata, the treatment *in loco morbi* consists in galvanisation, transverse and longitudinal, of the head; whilst the applications *in loco symptomatis* comprise labile galvanisation of the lips, tongue, and throat. The latter is best carried out as follows:—a large plate (anode) is fixed to the back of the neck, as high up as possible, whilst the lips, tongue, soft palate, and pharynx, are dabbed with a sponge fixed to a long electrode (fig. 71). The current is to be made as strong as the patient can endure it and applied for 3—5 minutes. Then (with a middle-sized disk electrode), the current is applied repeatedly to the sides of the throat externally so as to provoke marked movements of deglutition.

ELECTRISATION OF THE SPINAL CORD.

The principles which govern the rational application of electricity to the cord have been laid down and illustrated elsewhere, (p. 154). The galvanic current is to be applied with large electrodes for 5—10 minutes at a time, either stabile over the seat of the lesion when the latter is localised in any one particular segment; or slowly labile when the whole length of the cord is diseased and has to be brought under the influence of the current. When the upper part of the cord is galvanised, it is advantageous to place one of the electrodes (medium size) in the subaural position, first on one side then on the other (Erb). Tender vertebral pressure-points must be specially sought for and subjected to the influence of the anode. The strength of the current must, in "irritable" conditions of the organ, be weak and cautiously increased beginning with 8—12 cells. Much stronger currents (30—60 cells) are frequently well borne in chronic cases. Very large electrodes must then be used.

Electrisation of the cord is often conveniently combined with electrisation of the limbs; one pole (large plate) rests over the cervical or lumbar enlargement, whilst the other (of large or middle size) is applied to the peripheral nerves and muscles involved. Galvanofaradisation may usefully be carried out on this plan, especially in cases of atrophic paralysis of spinal origin.

With reference to the choice of the active pole, it is best to use the anode only in irritable conditions; the kathode in the more atonic ones, or better still the two poles successively whereby the so-called catalytic influence of the current appears to be enhanced.

Simple faradisation of the cord by means of moist electrodes has been recommended by some observers, as well as cutaneous faradisation over the whole length of the vertebral column, either alone, or in conjunction or alternation with, the galvanic treatment.

Under the name of *myelitis* are comprised various inflammatory states of the spinal cord, in which the lesion has the common characteristic of not being confined to any of its systems of fibres or cells. The prognosis will accordingly vary greatly from case to case. The localisation of the lesion forms the chief indication in the electrotherapeutics of chronic myelitis, (the acute stages of the disease do not present indications for this treatment); the current must be made to act as directly as possible upon the diseased structure (figs. 84 and 88). Paralysis of the legs, or *paraplegia* is the most common and striking symptom of myelitis; when the latter is transverse, that is occupies a whole segment of the cord, there also is anæsthesia, with vesical and rectal disturbances.

A strong application (20—40 cells) may usually be made daily, for five minutes, according to the rules first given. When the bladder is involved, voltaic alternatives or galvanofaradisation, with the electrodes placed on the sacrum and over the pubes may advantageously be administered during a couple of minutes. Galvanisation (stabile or gently labile) of the paralysed nerves and muscles may be combined with the central application, three or four minutes to each limb.

Chronic *meningitis* may beneficially be treated on the same plan, in addition to counterirritation by cutaneous faradisation of the back or the actual cautery. Much good may result from electrisation in cases of arrested vertebral caries where the spinal functions are interfered with.

In *tabes dorsalis* (locomotor ataxia) the use of electricity is sometimes attended with palliative results, either with reference to the rate of progress of the disease or to some of the numerous concomitant symptoms. The strength and length of the applications should be carefully adapted to the tolerance of each individual. They should be made daily and persevered in for many months.

As a rule moderate currents (15—30 cells) should be applied for five minutes along the spine. Erb recommends especially the subaural position for one of the electrodes, whilst the other is slowly moved along the spine. To this he adds labile applications to the legs, the anode resting on the sacrum, (or on tender spots, if such exist); and stabile applications for the lancinating pains, the kathode resting on the peripheral seat of pain, the anode over the spinal origin of the suffering nerve.

Rumpf has lately published some suggestive cases, where the daily cutaneous faradisation of the trunk and arms for five or six minutes proved very beneficial. The current must not be too strong and painful, and the wire brush passed over the surface rather firmly and slowly.

The treatment of *spastic paralysis* (lateral sclerosis) consists in moderate to strong galvanisation of the cord, and labile applications to the legs as just described. Duration about five minutes for the cord and as much for each leg; to be repeated daily.

In the states included under the name of *anterior poliomyelitis* (atrophic spinal paralysis, infantile paralysis), and where the lesion consists in an inflammatory change of the anterior horns of the spinal cord, the current should be directed both upon the cord and the degenerated muscles. The prognosis is more favourable in adults than in children; in the latter there usually is actual destruction of certain groups of the spinal motor cells, leading to irreparable atrophy of the corresponding muscles. Still in presence of the gravity of the evil, and the impossibility there is of knowing from the first how deep the lesion is, the duty of the physician is to insist upon the necessity of a systematic treatment, continued for a year or more if there is the least evidence of improvement.

The method of applying the galvanic (or better still the galvano-faradic) current is to fix a large plate electrode (anode) over the diseased portion of the cord (lower cervical vertebræ when the arm, lower dorsal and upper lumbar vertebræ when the leg, is affected), and with the other (a middle-sized well padded disk) to sponge over the whole extremity, with a number of cells sufficient to produce redness of the skin and muscular contractions. Each limb may be so treated for five minutes daily. Hot water with salt may be advantageously used to moisten the electrodes. It is most important that the affected limbs should be kept warm; for this purpose massage and frictions (before a good fire in the winter) repeated two or three times a day, and passive movements of the joints are essential. The part should be kept day and night covered with a cat's skin, and active systematic exercising of the weak muscles encouraged.

In true *progressive muscular atrophy* the results of any treatment are rarely even palliative. The galvanic current may be applied to the wasted muscles and to their spinal centres, by applying a large plate electrode over the latter and passing a middle-sized disk electrode over the periphery. The current should not be strong, the number of cells being chosen so as not to produce any well-marked muscular contractions in the part. The poles should be occasionally changed during the sitting; the latter not to be so protracted as to exhaust the diseased structures.

In the disorders known as spinal *irritation*, *neurasthenia*, and the "*railway-spine*," the galvanic treatment of the cord must be mild. The current may be applied daily, or every other day, according to the methods already given, for about five minutes. Diligent search must be made for spots tender to pressure; a weak anodic influence of eight or ten minutes duration to each of them often acts beneficially. Cephalic and subaural galvanisation also is often indicated in such cases; and the galvanic treatment supplemented with general faradisation, especially of the muscles of the back.

FUNCTIONAL CEREBRO-SPINAL DISEASES.

THE electrical treatment of the more general functional nerve diseases, or neuroses, has to fulfil, according to the nature and peculiarities of each case, one or more of the three indications: (1) removal of the cause, (peripheral irritations, etc.); (2) relief of the prominent symptoms (paralysis, spasm, pain, etc.); (3) restoration of the disturbed organ (molecular, nutritive changes in the nerve centres).

The first two indications require methods of procedure which the reader will readily gather from what has already been said concerning the electrotherapeutics of nervous diseases; with reference to the third, he must bear in mind that general neuroses require general treatment, that is to say, in the case of electricity, applications to the whole cerebrospinal axis, and applications to the whole of the periphery so as to act both directly and reflexly upon the nutrition of the whole of the nervous centres, (central galvanisation, general faradisation).

The treatment of cerebro-spinal *neurasthenia* has been described along with that of other affections of the brain and spinal cord. General electrisation may do much good in *hypochondriasis*, especially when the abdominal organs appear to be at fault. Here galvano-faradisation is indicated with very large electrodes applied to the abdomen and lower dorsal spine.

In all neuroses where there is evidence of emotional disturbance no treatment is of any avail which is not supported by appropriate moral influences. This is especially the case in *hysteria* where a complete removal of the patient from all her friends and surroundings is usually a *sine-qua-non* condition. The Weir Mitchell plan of treatment, which has been successfully applied in this country by Dr. Playfair to many cases of hysteria (especially to such where inanition played a part), consists in seclusion, general faradisation, massage and "over-feeding."

It is a mistake to assume that all the symptoms of the hysterical neurosis always yield speedily to electricity. When they do so, as is often the case for aphonia, the effect may be due to the sensory and psychical influence of the application. Hysterical spasms and paraplegia are by no means inviting objects to the scientific electrotherapeutist who will, if called upon to interfere, rely mainly upon the general measures which must ever form the leading part of the treatment. Nor will he ever fall into the fatal error of resorting to violent methods in the hope of accomplishing one of those startling cures, which though frequent in hysterical derangements, are to be obtained rather by appropriate psychical influences. Experience has shown that over-electrisation in functional nerve disease may do positive harm to the patient who thereby loses confidence in the treatment, and thereby shuts one possible issue to his sufferings.

The local symptomatic treatment of hysterical anæsthesia, neuralgia, paralysis, etc., will be conducted on essentially the same lines of action as those previously described in the general account given of those symptoms.

In *chorea* and *epilepsy* we find occasionally symptomatic indications for electrical treatment (paralysis, etc.), which has to be conducted on the principles stated previously (p. 165, etc.) The recent views concerning the motor functions of the cerebral cortex, and its implication in certain choreiform and epileptiform phenomena, lead one to think that in those cases at least where the usual medication fails cephalic galvanisation should be tried. Some examples indeed, reported on good authorities, of beneficial results obtained by it, have already been published. The rationale being that the current exerts some influence upon the nutrition of the altered nervous tissue, the mode of application will be that described on page 175. One electrode is to be placed over the convolution in which the morbid process has been localised, whilst the other rests upon the nape of the neck.

The galvanic current is of great use in *exophthalmic goitre*, though in advanced cases we may scarcely hope for a cure. It should be applied perseveringly for one or two months consecutively, and several such courses be resorted to at intervals. The applications should comprise galvanisation of the brain, from the eyes and forehead to the occipital region, and through the mastoid processes; subaural galvanisation; and galvanisation of the upper part of the cord (labile, with indifferent electrode on the epigastrium). The currents should be weak, especially at first, and the sittings short; gradually increasing as tolerance is established; the manipulations must be conducted on the general rules given above. In a recent case of exophthalmic goitre I obtained marked and speedy results from voltaic alternatives with a rather strong current, which the patient happened to bear exceedingly well. The electrodes were applied according to the subaural method (p. 164).

In *paralysis agitans* and *tetanus*, electrisation has yielded but slight palliative results. *Diabetes mellitus* and *insipidus*, the latter especially, have been treated with galvanisation of the nervous centres. Whether the results reported to have been obtained are real and durable remains to be shown by further attempts in that direction.

Writer's cramp. This name is applied to the most conspicuous of a group of what might be called "professional functional neuropathies." These have their mode of origin in common, viz., the exercise of certain groups of muscles in the performance of certain co-ordinated movements of later acquisition, such as writing, playing musical instruments, telegraphing and the like. Writer's cramp is also known as 'scrivener's palsy,' a name which shows that the spastic form of the disease is not the only one. In some cases paresis is the main symptom, in others again tremors and paræsthesiæ are conspicuous.

The nature and localisation of the lesion are both matters of speculation. The treatment consists of rest—absolute—from the muscular actions, which give rise to the symptom, electrisation, and a combination of gymnastics and massage. The latter being still almost universally ignored in this country, I shall give here an account of the German method; using Ross' words, "the gymnastic exercises consist of both active and passive movements. In the active form, the patient is instructed to execute, three or four times a day, a series of vigorous movements with the affected extremity, the hand being opened and closed in quick succession. The number of these movements, and consequently the duration of each exercise is progressively increased until a duration of half an hour is attained for each sitting. In the passive movements the operator produces forcible traction, three or four times a day, upon each of the affected muscles separately, in the direction of its length. This is the most delicate part of the treatment, inasmuch as, if too little strength is employed the cure is delayed, if too much the disorder is aggravated. When the spasm is notably diminished, which usually occurs in a short time, the patient is encouraged to take slow and graduated exercises in writing. The operator practises daily massage of the affected extremity, particular stress being laid upon percussion of the affected muscles with the ulnar border of the hand." (This method usually ascribed to Wolff who demonstrated it, and obtained brilliant successes in Paris, appears to have been originated by Dr. Schott, of Nauheim. (*Cf. Neurologisches Centralblatt*, April, 1882).

The electrical applications should consist of labile kathodic galvanisation of the brachial plexus, and of the muscles and nerves of the arm, the anode resting over the cervical enlargement, for ten minutes daily. Weak currents; fifteen or twenty cells will usually be found sufficient. The active gymnastics mentioned above may be usefully carried out during the electrisation.

Migraine or *Sick-headache*. This ailment is to be considered as a manifestation of a neurotic constitution rather than as a local nervous disturbance; and consequently its treatment will have palliative rather than curative results. At the same time electricity should not be applied symptomatically with a view to cut short the individual attacks, but in a systematic and continuous manner for a long period in order to improve the nutrition of the abnormal nerve centres.

The application should be made daily:—1st, through the head, (p. 175) one pole resting on the subaural position whilst the other is applied to the temple and mastoid process on the opposite side, according to the directions given above; 2nd, from the nape of the neck to the epigastrium, moderate current (12 to 18 cells) for 6 to 8 minutes. During the attacks, and also in the intervals, a faradic current may be applied through the head (from forehead to occiput) by means of two large well padded electrodes, or better still with the moistened hand of the operator used as one of the electrodes, the current passing through the body of the latter who holds the real electrode in his other hand.

Holst's polar method is more theoretical than practical. When pallor is present (angiospastic attack, he places the anode in the subaural position and gradually makes a current which he allows to flow for about 5 minutes and gradually removes. When hyperæmia is manifest (angioparalytic attack) he seeks to excite the sympathetic by placing the kathode in the same position, and interrupting repeatedly the current.

DISEASES OF THE PERIPHERAL NERVES.

The conditions which call for electrical treatment of peripheral nerves are (besides functional disturbances) neuritis, perineuritis, injuries, degeneration. The symptoms of these morbid states are paralysis, pain and trophic changes in the area of distribution. After what has been said of the rationale of electrisation in paralysis little remains to be added here of a general nature. The catalytic influence of the current *in loco morbi* and its exciting and modifying effects both above and below the seat of lesion will obviously fulfil the most important indications; and the methods of electrisation will not offer many difficulties to any one conversant with the relative position of the nerves affected and the laws of current-diffusion. The principal motor points will be found illustrated in the diagrams appended to this chapter.

Oculomotor, Trochlear and Abducens.—Paralyses in the district of the third, fourth and sixth pairs of cerebral nerves are sometimes of central origin (bulbar, "ophthalmoplegia externa"), more commonly of peripheral nature (syphilitic, rheumatic). They also occur in the course of tabes or locomotor ataxy. The prognosis will obviously depend upon the cause.

The galvanic anode is to be fixed to the nape of the neck, and the kathode (a small electrode, well covered with soft sponge) applied upon the closed eyelid, over the point of insertion of each of the paralysed muscles successively, a certain degree of firm pressure being exerted. The current (6 to 12 cells) must be strong enough to produce distinct twitches in the orbicularis. The current is allowed to flow for one or two minutes through each muscle, and then a few interruptions are made, or if the patient objects to them the electrode is passed over the closed eyelid for the same period. Faradisation has been used, but is inferior to galvanisation.—I am in the habit of using the two currents combined (see p. 163), a weak faradic current being made to pass simultaneously with the galvanic used as just described.

Facial nerve.—Paralysis of the seventh nerve is usually due to exposure to cold, and is then supposed to be due to a perineuritis, with effusion into the sheath. The nerve fibres are thus compressed, and paralysis ensues, with or without degeneration (p. 127) according to the degree of pressure to which they are submitted. Among other causes of "facial hemiplegia" of peripheral origin, may be mentioned injury to or disease of the nerve as it passes through the pons along the base of the skull, and in the bony canal. In the disease known as bulbar or labio-glosso-laryngeal paralysis certain nuclei of the medulla oblongata degenerate, among them that presiding over the movements of the labial muscles. In the usual hemiplegia from cerebral hæmorrhage the facial paralysis, which occurs on the side opposite to the lesion, does not extend higher than the zygomatics. In peripheral paralyses on the other hand (including alternate hemiplegia from disease in the pons) the whole of the facial district is involved.

We have seen in the chapter on electrodiagnosis that facial paralysis (from exposure to cold, hence called "rheumatic") offers three varieties according to the depth of the lesion.—In the lighter cases the reactions are normal and the nerve is injured only *quoad* its functions as an organ of transmission of voluntary stimuli. Neither it nor the muscles undergo degenerative changes; and the recovery is accordingly rapid (about 3 weeks).—When the reactions present phenomena of the middle form of RD (p. 130), and prove that the trophic influence of the central nuclei on muscles is cut off, the progress is slower, about two months elapsing before recovery.—In the severer type, where nerve and muscles are degenerated, the progress is slower (six months or longer) and may end in a condition of permanent impairment. Facial paralysis thus yields a typical picture of RD in its various stages of development and an excellent example of the prognostic value of an electrical investigation.

The treatment of facial paralysis consists of a thorough labile galvanisation of the nerve, its branches and every muscle it supplies, first with the kathode, then with the anode, whilst the other pole rests upon the point of emergence of the nerve (*i.e.* well pressed in behind and below the ear). The current should be as strong as the patient can bear (8-12 cells), and applied for about five minutes with medium sized electrodes. The effect of the treatment will obviously be greater when the time for the natural process of regeneration has arrived (see diagrams, figs. 79, etc.) Its immediate effect is to restore some tonus to the muscles, and to remove the uncomfortable "stiff" feeling of which many patients complain.

Laryngeal nerves.—Electrisation of the larynx is frequently resorted to in various forms of paresis of its muscles. It is by no means necessary to resort to the difficult and unpleasant method of intralaryngeal faradisation which is usually described by specialists, as galvanisation applied through the neck is quite sufficient to excite the parts. For this purpose a plate is fixed to the nape of the neck, and the skin over the larynx and both sides of the trachea vigorously stroked with a medium sized electrode; then a number of voltaic alternatives are to be made whilst the latter remains stationary over the larynx and over the vagus. Current as strong as can be borne without too much discomfort (about 10 cells). Galvanofaradisation applied in this manner is very effectual. Hysterical aphonia is usually amenable to any form of electrisation; the current must be strong and suddenly applied so as to produce a psychical effect which often finds vent in a cry.[*]

Phrenic nerves.—*Artificial respiration.* Electrisation of the phrenics is rarely resorted to in actual disease, but constitutes an important factor in the procedures for restoring life in asphyxia, etc. The simplest plan is to press a medium sized electrode well behind the the posterior edge of the sternomastoid (see figure of motor-points)

---

[*] I have just seen a case of six years' standing, where strong faradisation failed to produce the desired effect. The voice returned however on the application of a powerful galvanic current with rapid interruptions.

on each side. The negative pole of a sufficiently powerful coil is connected with these electrodes by means of a branched rheophore; whilst the positive rests on the epigastrium. A very strong current is made for one or two seconds and interrupted for same period. The arms and head are kept fixed during excitation; and pressure made upon the thorax to assist expiration during the intervals.

Nerves of the arm.—The roots and trunks which form the brachial plexus are easily submitted to electrical influences by a proper disposition of the electrodes, which when placed over the cervical vertebræ, in the supra clavicular fossa (see plate of motor points), and in the axilla secure a considerable current density in the upper segments of those nerves. It must be remembered that the first dorsal and eighth cervical root supply the flexor of the hand and fingers, and small muscles of the hand chiefly; the seventh cervical (in man) the extensors chiefly; whilst the supinators and flexors of the arm seem to be mainly innervated by the fifth and sixth root, along with the deltoid, and scapular muscles. The triceps receives branches from the lower three cervical roots, a fact which explains the rarity of its complete and isolated paralysis (*cf.* p. 137). Disease of the motor roots can therefore be distinguished from lesions of the nerve cords and trunks below the plexus by the fact of their involving physiological groups of muscles, like poliomyelitis, and not anatomical groups or groups supplied by the same nerve.

The conditions which call for electrical treatment in nervous disorders of the arm are numerous, including chiefly neuritis, traumatisms of various kinds, the effects of cold, of rheumatism, etc. As usual applications *in loco morbi* have to be made with sufficiently large and well disposed electrodes (*cf.* p. 153), the current (of moderate strength) averaging from 15-25 cells, and lasting from five to ten minutes. Excitations are to be made above the point of lesion by means of faradisation or galvanic shocks, and the whole muscular province innervated by the diseased organs subjected to a thorough labile galvanic or galvanofaradic application for five minutes or more with both poles sucessively and a medium sized electrode. During this part of the treatment the other electrode may be made to rest over the seat of the lesion and thus the two indications fulfilled at the same time.

Nerves of the Leg.—The same general remarks apply to the electrical treatment of the nerves of the leg as to those of the arm. The position of the plexuses, lumbar and sacral, however make it more difficult to reach them with currents of sufficient density. Very large electrodes are to be used with correspondingly powerful currents (25 to 40 cells or more especially in the case of fat people). Galvanofaradisation with voltaic alternatives, is indicated here; ordinary faradisation is of but little avail, except when applied per rectum (p. 192). The peripheral treatment should be conducted as in the case of the arm; the number of cells required being generally larger on account of the higher resistance of the tissues. The spinal localisations of the motor centres of the lower extremities will be readily understood by a reference to page 135, and to diagrams of the roots and plexuses and nerve trunks such as are usually given in textbooks of anatomy. The course of the great nerves (sciatic, crural, etc.) need not be indicated here. The motor points of the several muscles will be found indicated in the figures at the end of this chapter.

TOXIC PARALYSES.

Lead paralysis.—The commonest form of lead paralysis is that affecting the extensors of the hand and fingers, (the supinators usually remaining intact), hence called wrist-drop. The pathogeny of the disease is still a question of debate. I have (*Lancet*, 1880) supported the hypothesis that it is of spinal origin, and consists in a functional disturbance of the grey matter, a poisoning of the cells, giving rise to an arrest of their functions, motor and trophic. Acting in accordance with this view I have been in the habit of applying the current from the cervical enlargement to the paralysed muscles.

A large electrode is fixed over the middle cervical vertebræ, and the upper surface of the forearm is subjected to a vigorous labile application of a current of 15 or 20 cells or more. The two poles are altered several times during the sitting.

The prognosis will depend on the depth of the neuro-muscular lesion, as judged by the degree of alteration of the normal response. RD is usually well marked in saturnine wrist drop; a fact which along with the escape of the supinators serves to distinguish it from pressure-paralysis of the musculo-spiral.

Diphtheritic paralysis.—Among the sequelæ of diphtheria occur paralyses of the soft palate and muscles of deglutition, as well as of other muscles of the trunk and extremities. These are amenable to galvanisation which is to be applied as described in the paragraph on Bulbar paralysis, and in those of peripheral paralyses.

Nervous disturbance *after acute febrile disease* will be treated by central and peripheral applications according to the principles laid down in previous paragraphs.

Syphilitic paralyses occur most commonly under the form of hemiplegia and paraplegia when the centres are affected. Among the peripheral nerves the oculomotor is most commonly affected. The first indication is of course the causal one; and here, as in all specific neuropathies, a very energetic treatment with mercury and iodide is of the utmost necessity.

The directions given previously for the symptomatic treatment of the various paralyses will have to be afterwards followed whenever the full restoration of innervation remains in abeyance in spite of the internal medication.

DISEASES OF THE EYE.

Good results are reported as having followed the application of electricity in a number of cases of disorders affecting the ocular structures anterior to the retina. The latter need not be alluded to in detail here; I shall confine myself to the statement that in further experiments on the subject the operator should act on the general principle that as the nutrition of the tissues is to be influenced, it is best to apply both galvanic poles in succession over the closed eyelid. The indifferent electrode should rest upon the nape of the neck or in the subaural position. Weak currents to be used at first, gradually increased (4-10 cells) for 4 to 8 minutes daily. The faradic current is said to be useful in some cases; it should be applied as just described only for a longer time (thirty minutes to one hour, once or oftener every day).—The eye electrode should be oval, somewhat concave, and covered with a layer of fine sponge; a flexible metal plate of appropriate shape and size is more convenient than the usual handle and disk electrode, as it may be kept in uniform apposition with the eye by means of a bandage.

Galvanisation usually brings about an improvement in *retinitis pigmentosa*.

Primary *atrophy of the optic nerve* appears to be unfortunately beyond the reach of electricity as of all other methods of treatment. Many patients however are anxious to try the effects of galvanism; they may safely be ordered to apply the current of three or four cells to each eye for 5-10 minutes daily, or twice a day, according to the method just described.

In *optic neuritis* and its sequelæ the current is of much greater value. In addition to the direct galvanisation of the eye, it is recommended that subaural galvanisation should be practised.

Ocular affections symptomatic of nervous disorders such as the choked disk of cerebral tumours and the atrophy of locomotor ataxy cannot be expected to yield to special treatment. In certain functional disturbances of the retina and optic nerve galvanisation and faradisation may legitimately be resorted to.

DISEASES OF THE EAR.

The application of electricity to the organ of hearing is a subject of historical, as much as of practical interest. It was through investigation of the acoustic reactions Brenner was led to establish his polar formula, and to lay down the rules for the unipolar investigation of motor nerves and for the unipolar treatment of disease. The violent controversies to which his publications gave rise are now a thing of the past even in Germany, the only country where scientific electrotherapeutics ever evoked a feeling of interest; but Erb has quite lately restated Brenner's views in the light of recent observations, and with the support of his authority in such matters.

We have seen (p. 117) that the healthy acoustic nerve reacts in a definite manner to anodic and kathodic excitations. Now in morbid conditions, quantitative or qualitative alterations are observed in this normal polar formula, which yield data both for diagnosis and for treatment. The symptoms which call for galvanic treatment are deafness and tinnitus, due respectively to anæsthesia and hyperæsthesia of the acoustic nerve.

In nervous *tinnitus* the reactions of the nerve are probably always abnormal; and the indications for the treatment are derived from the respective influence of the anode and of the kathode on the subjective sounds, that pole being applied to the ear under whose influence the noises are found to diminish. In certain cases complete silence is produced, especially by the anode, for a longer or shorter time: and here the prognosis would appear to be more favourable than where the effect is less marked.

Many of the ill successes which have in the hands of aurists attended the galvanisation of the acoustic nerve would seem to be due to the very defective methods employed. In order to make the diminution of the tinnitus persist after the removal of the electrode it is indispensable that the latter should be raised only when the current has been reduced to zero by imperceptible degrees. Now this is an impossibility with the usual batteries without the intercalation of a rheostat, or some improvised resistance such as a moistened string (p. 11) in the circuit, and the use of a collector taking the cells by twos at most. When both ears are affected, and indeed in all cases in order to diminish the giddiness attending the circulation of a current through the brain, it is best to use a bifurcated rheophore connecting an electrode placed on each ear with the active pole, while the indifferent electrode rests as usual on the sternum or some other distant part of the body.

The strength of the current should be such as to check the tinnitus if possible without producing much giddiness; and the application be protracted to ten minutes or even longer. When the effect of the galvanisation lasts for a few hours, it would perhaps be advantageous to repeat the process oftener than once a day as in the case of neuralgia, but I am not aware that this plan has yet been tried.

In such cases, and they are by no means unfrequent, where the strictly electrotonic method fails to secure permanent results, a further attempt must be made with a view to catalytic effects, viz., the nerve must be submitted to the effects of the two poles successively. The best plan is to begin with the kathode, and pass to the anode by a sudden reversal of the current, which is finally reduced gradually to zero as before.

The plan for the treatment of nervous *deafness* consists in excitations of the nerve with a galvanic current in which interruptions or voltaic alternatives are made. The precautions for gradual diminution of the current are of course unnecessary; but the mode of application otherwise remains the same as for tinnitus. When the latter coexists with deafness, and is relieved by the current, applied according to the method described above, both are usually found to disappear at the same time.

NEUROSES OF THE HEART AND LUNGS.

Asthma.—From what has already been done by electricity in cases of nervous asthma there appears to be every ground for more extensive efforts in that direction. One plan consists in the energetic faradisation of the vagi below the angle of the jaw, twice a day for about 20 minutes. Galvanisation of the upper part of the cord and sides of neck has also been employed. The most ready and complete method appears to me to fix a medium sized electrode on the occipital region, and apply the galvanofaradic current from the subaural region down to the sternum for 6 to 8 minutes on each side. The number of cells should be graduated according to the tolerance of the patient.

Angina Pectoris.—Duchenne employed with some success the induced current applying it by means of the wire brush to the præcordial region as a powerful derivative. The galvanic current may be advantageously used also, the anode rests on the præcordial region whilst the cord from the occipital region to the interscapular region is subjected to the kathode, as well as the sympathetic and vagus on either side of the neck. The current should be weak and the sittings short at first, and only gradually increased.

Nervous palpitations are sometimes amenable to galvanic treatment according to the method just indicated. The physiological effect of the galvanic current on the muscular fibres of the heart has been mentioned previously (p. 114).

DISEASES OF THE ABDOMINAL ORGANS.

A NUMBER of disorders of the digestive organs present indications for electrical treatment, though it must be confessed that this field of therapeutical activity has hitherto been chiefly occupied and cultivated by quacks. Abdominal neuropathies, congestion and atony indeed form more hopeful subjects for electrisation than many of the disorders of the central nervous system in which the method is too commonly supposed to be peculiarly adapted.

We have seen (page 153 *ff.*) the principles which ought to guide every effort to bring deep organs under the influence of the current; nothing but failure can follow departures from the rules derived from the physical conditions under which we are acting in abdominal electrisation. Very large electrodes (p. 90) and strong currents are necessary. Hence it follows that the ordinary medical induction coils are of but little use here on account of the very small quantity of electricity they yield, and faradisation of the deeper organs in order to be effectual must be made with large coils of thick wire.[*] The ordinary medical pocket induction ap-

---

[*] I have had an instrument made for this purpose which admirably fulfils its object. It consists of a primary coil of about 60 yards of *very* thick wire (2 mm. diameter) fed by two large Leclanché cells. The current is graduated by means

paratus is thus to be rejected, and in all cases it will be found advisable in order to increase the effect of the induced current to apply it as described in the paragraph on galvanofaradisation (p. 163 *cf.* p. 111), combining it with a galvanic current. Voltaic alternatives of the galvanofaradic current form the most powerful excitant in electrotherapeutics, and will be found useful whenever the deep abdominal structures require to be influenced.

Whenever muscles of organic life are to be excited it is best to use the labile method; or the stabile with rather long periods of interruption.

The stomach may be faradised internally by means of a wire introduced through an india-rubber tube, and the rectum by means of the electrode shown at fig. 69. The former process can always be dispensed with, when a sufficiently powerful coil is at hand; the latter is very simple and effectual, and easily tolerated when the electrode is not pushed against the branches of the sacral plexus.

Dyspepsia is a symptom of various conditions of the stomach in some of which electrisation will be found useful either as a palliative or adjunct, or as a causal treatment. This will be the case when the functions of the gastric nervous and muscular structures are at fault. Moreover general faradisation, or perhaps the electric bath, are indicated by the debility of the whole system which often accompanies dyspepsia. When the symptoms of *gastralgia* prevail the best method is to fix a large electrode over the region of the stomach and to apply the other (slowly labile) along the dorsal portion of the cord. A strong galvanic current may thus be used for 8 to 10 minutes. There are no definite indications as to the polarity of the electrode resting over the stomach, the effect of both may be tried successively.

When there is *paresis* of the muscular walls, or *dilatation*, galvanofaradisation will prove the most efficient method, and may well be combined with the process known as washing out of the stomach. A very large electrode is fixed to the back, and another applied successively to all the points of the surface over the dilated organ. The application should consist of a series of excitations 30 or 40 in number, each lasting 15 to 20 seconds, with short intervals, and voltaic alternatives.—Certain cases of *flatulence* and altered secretion, especially those which are not amenable to ordinary medication and hygiene should be submitted to a similar treatment in the hope of checking the morbid process by rousing up the nervous vigour of the organ.—*Vomiting* of a purely functional and idiopathic description is amenable to moderate galvanisation of the vagus and stomach. For this purpose a large plate is to be fixed to the epigastrium, whilst the other electrode, of medium size, is made to rest on the occipital region for 3 to 5 minutes, then in the subaural position and along the trachea on each side for the same

of the moveable core, and when applied to the abdomen through two very large plates (thoroughly moistened in hot salt and water) produces powerful contractions of the whole portion included. Notwithstanding the great diffusion of the current the quantity of electricity it conveys is sufficient to excite the whole of the tissues *en masse*.

period. Moderate current (10 to 20 cells), both poles may be tried in succession.

Constipation is a symptom as common as troublesome, but which in many cases yields to a rational application of electricity. The most effectual method is the galvanofaradisation of the whole abdomen. For this purpose a very large plate is fixed to the back, whilst another is made to rest first upon the umbilical region, then carried round along the whole course of the colon. Strong currents for 8 or 10 minutes daily, with numerous voltaic alternatives. When the subject is very fat a good deal of pressure should be exerted upon the electrodes so as to diminish the distance between them and the viscera.

Faradisation alone may be used, or galvanisation, in the same way as galvanofaradisation. It is well, however, when the former only is applied, to use a rectal electrode (see fig. 69) pushed well up into the bowel, and a medium sized disk electrode very firmly pressed upon the abdominal walls.

Similar applications, only stronger, longer, and more frequently repeated often prove highly successful in *obstruction* Cases are on record where copious evacuations took place immediately after the first excitation of the intestines after all other means had failed. Whether obstruction from other causes than fæcal accumulations can be relieved is a point not clearly made out, though well authenticated instances, however, *e.g.*, of invagination, encourage us to further experiments which appear to be at least quite free from danger.

Flatulence and *meteorismus* when not symptoms accompanying the states just described, may call for treatment, which is to be conducted on the same lines as that indicated in simple constipation.

Certain painful conditions of the abdominal sympathetic and its branches—*abdominal neuralgia, enteralgia, dry colic*, etc., are amenable to galvanisation. The applications should consist of stabile currents from the back to the abdominal walls. Large plates, 30 to 40 cells, 10 minutes; with a few voltaic alternatives. *Lead colic* is also amenable to faradisation (see "Constipation").

Congestion of the portal system, especially when accompanied with cerebral depression, and general atony of the digestive organs of neurotic nature may well be subjected to the effects of galvanofaradisation.

Certain *enlargements of the spleen* have frequently been proved to be amenable to electrisation, galvanic or faradic. Two large electrodes are to be placed so as to include the organ and moderate or strong currents, with interruptions used for ten minutes daily. It is said that in old cases of ague the other symptoms are alleviated as the organ is reduced in size; and that after each electrisation marked changes in the relative number of white and red blood corpuscles can be demonstrated by the hæmocytometer.

As a point of physiological interest, if not of great therapeutical import, may finally be noted the effects of faradisation on *ascites* from obstructed circulation. The rhythmatical exercise of the abdominal walls increases the absorption of the liquid through the lymphatic stomata and the urine becomes more abundant as the ascitic effusion diminishes. The plan deserves more extensive trials than it yet has had.

DISEASES OF THE MUSCLES AND JOINTS.

Simple muscular atrophy, such as is often met with in many surgical cases, is readily amenable to faradisation. Each of the affected muscles must be made to contract thoroughly (for a few minutes daily) by applying one of the electrodes on its nerve or motor point, whilst the other is carried over its whole surface. The same treatment is useful after such operations as resection of joints, etc. The stiffness and weakness which sometimes remain after fractures, contusions and other *injuries* of the bones, joints, ligaments, tendons or muscles, are beneficially influenced by a more or less energetic treatment, either with the galvanic or the faradic current. I usually apply both together (see galvanofaradisation p. 163). The applications should consist of a thorough electrisation at the seat of the injury, as well as of the whole muscular masses connected with it. *Sprains* are especially amenable to this treatment; and it is not rare to see patients who had painfully hobbled into the room, walk out with comparative ease. In many cases it is the pain rather than the real damage to the parts which prevents the use of the limb, and this is most effectually removed by the application of galvanofaradisation. I have used it with success to quite recent cases as well as to very chronic ones; it is prudent however when there is much effusion and inflammation to use at first nothing but mild and protracted faradisation at the seat of injury, and wait a few days before passing to more vigorous applications.

Muscular rheumatism is often amenable to cutaneous faradisation over the seat of pain (moderate currents, two or three minutes); or to faradisation of the muscles themselves. The latter method is best carried out by placing two well-moistened electrodes of good size over the muscular masses and begin with moderate currents which are gradually brought up to the maximum strength endurable. Duration five minutes, or more, to be repeated more than once a day if necessary. Galvanofaradisation may with advantage be substituted for simple faradisation.

The galvanic current may also be applied to the painful parts. The electrodes are applied according to the principles above mentioned (p. 153) and a current, moderate to strong, passed for a few minutes, and the sitting concluded with a series of interruptions or voltaic alternatives so as to excite the muscular tissue to contraction.

Chronic articular rheumatism proves often very obstinate to every treatment, but occasionally yields to persevering electrisation. Here again, stabile galvanofaradisation of the joint (moderate currents, 10-15 minutes), is perhaps the best; after which labile applications are to be made, extending some distance above and below it so as to excite all the muscles in relation with it. A rather unusual method of applying the galvanic current has lately been extolled by Prof. Seeligmüller, viz., by means of the wire brush. The anode (a large moistened electrode) rests upon any part of the body, whilst the kathode (a brush with stiff wires) is firmly applied to several points over the seat of pain, resting on each for a few seconds. The current (15-20 cells) must be strong enough to produce a number of small eschars wherever the brush is applied. This method is obviously very painful, but the results obtained are described as so rapid and marked that the patients gladly submit themselves to it.

Arthritis deformans when advanced admits of palliative treatment only. On the supposition that it is of nervous origin, it is best to fix the anode over the spinal origin of the nerves supplying the diseased part, and submit the latter to slowly labile applications of the kathode, including the muscles connected with the articulations. Moderate currents, ten minutes for each limb. In addition to this, spinal galvanisation may be practised according to the subaural method (p. 179) on the theory of the neurotrophic nature of the complaint. The treatment should be persevered in for a considerable period.

DISEASES OF THE GENITO-URINARY ORGANS.

Paralysis of the bladder is a frequent symptom in diseases of the nervous centres, and forms an important indication for local treatment. In many cases electricity will be found very useful; but whenever the cord is gravely affected (myelitis, transverse or diffuse, etc.) it naturally fails to relieve.

The best method is to apply galvanofaradic currents by means of very large electrodes from above the pubes to the perineum and to the sacrum with frequent reversals or voltaic alternatives by means of the apparatus described at page 96, but the galvanic current alone is to be applied in the same way in the absence of the necessary arrangements. The currents should be strong or very strong, and applied for 8-10 minutes. When the paralysis is of central origin we submit also the lumbar enlargement of the cord, and in some cases the whole cord, to the electrical influence (see page 178). Internal electrisation of the bladder may be also resorted to if external applications are not successful.

Enuresis nocturna is frequently amenable to treatment; cases of many years standing often yield with extraordinary rapidity to electrisation of the cord and bladder. It is best to begin with external applications similar to those described in the preceding paragraph, but milder; and if results are not speedily obtained to try the internal galvanisation or faradisation of the bladder. For this purpose a gum catheter with a metallic tip connected with one pole of the battery is introduced into the bladder, and the circuit closed by means of an ordinary plate electrode over the abdomen. When faradism is so applied the currents must be strong, and the primary (thick wire) coil used; when galvanism, the currents must be weak, and in order to avoid producing eschars on the walls of the bladder, the latter should be partially distended with tepid water. The duration of faradic applications may extend to five minutes; but when galvanism is used the current should not be allowed to flow continuously for more than a few seconds at a time, or voltaic alternatives should be made frequently. Weak currents (8-12 cells), for three minutes.

Impotence and spermatorrhœa, when of functional origin, have often been shown to be amenable to electrical treatment. In certain cases the moral effect is not to be lost sight of ; these affections are peculiarly prone to upset the mental equilibrium of patients, and much firmness and discrimination are always required on the part of the physician in their management. With reference to the mode of applying the current, the easiest and most effective, is to fix a large plate over the lumbar enlargement of the cord, and by means of a disk-electrode covered with a layer of soft sponge submit the perineum, groins, internal surface of the thighs and genitals, to a thorough labile application, either of galvanism alone, or better of galvanofaradisation, with frequent voltaic alternatives. The current must be as strong as can be borne, and the duration 8-10 minutes. The plate is then conveyed to the abdomen, and a similar labile application made to the sacrum over the cauda equina for 3-5 minutes.

In certain cases, chiefly of simple spermatorrhœa, this plan may prove too energetic, and stabile or slowly labile galvanisation of the cord, and chiefly of the lumbar enlargement, is to be preferred. The anode may be applied to the perineum, and the kathode used to the back ; moderate currents every day or other day for 8-10 minutes. The vesiculæ seminales may easily be faradised by means of a rectal electrode (fig. 69) ; or a weak galvanic current (8-10 cells) applied by means of the urethral electrode (fig. 71) to the often hyperæsthetic orifices of the ducts for 2 minutes. These local applications may be made conjointly with the more general ones just described.

The efficacy of electrisation in *amenorrhœa* of functional origin is now universally recognized. It appears to act chiefly by reflex action, and need not necessarily be applied to the uterus itself. Energetic faradisation or galvanisation, better still galvanofaradisation, from lumbar to suprapubic region, (with large plate electrodes for 10 minutes daily, especially at the time the menses should appear) should be tried first, in cases where intrauterine electrisation seems objectionable. Otherwise a uterine electrode connected with the negative pole of the galvanic battery (fig. 70) is introduced and the circuit closed, by placing the anode upon the abdomen or back. Weak current (about 8 m.a. according to

Dixon Mann) for 10-15 minutes every other day. The latter method is useful in certain cases of *menorrhagia*, whilst nervous *dysmenorrhœa* may be relieved or cured by the internal, or external applications just described, and which should be, as usual, carried out between the menstrual periods.

Electricity has been highly recommended by several gynæcologists for the treatment of *chronic metritis* and certain *uterine displacements*. The rationale is to act on the muscular elements of the organ itself and its surroundings. For this purpose either internal faradisation or galvanisation has been resorted to. The combined use of both currents with voltaic alternatives seems to me more promising in the accomplishment of the aim in view. The currents should be moderate, and applied about three times a week for about eight minutes.

I have frequently had the opportunity in the out-patients' room to observe the beneficial effects of the current on the pains and sense of weariness or discomfort depending upon various ovarian or uterine derangements. The palliative results so obtained may be more or less permanent; but generally speaking, electricity seems to deserve a much more extensive trial in uterine disorders than it has hitherto had.

In midwifery practice faradisation of the uterus (one pole on the cervix, the other on the abdomen) is useful to excite the contractions in cases of insufficiency during labour or of hæmorrhage afterwards; the advantage of this method over the exhibition of ergot is that the effects of electricity are more complete and imitate nature by being intermittent.

Excitation of the breasts by faradisation has been found effectual in cases of deficient milk secretion. Frequent applications of a quarter of an hour, with rather strong currents, should be made.

DISEASES OF THE SKIN (CUTANEOUS TROPHONEUROSES).

It has become generally recognised of late years, that in a number of cases of cutaneous disorders, the nervous system is primarily at fault. As a sensory organ the skin is brought into a close physiological relationship with the nerve centres; whilst embryologically it has a common starting point with them, in the external germinative layer.

Numerous lesions of the nerve-centres and nerve trunks, are followed by well marked cutaneous trophoneuroses, such as decubitus acutus, perforating ulcer, glossy skin, various eruptions, alterations in the hair and nails, etc.

The pathology of many neurotrophic skin diseases, however, is still obscure; but it must not be forgotten that the absence of post-mortem changes in the nerve centres by no means excludes their participation in the production of peripheral morbid processes. For as I have shown, some years ago,* with reference to the neuromotor apparatus, we may as well assume that a *functional* disturbance of the trophic centres will produce a visible manifestation—viz., an *organic* lesion—at the periphery, as that a dynamic alteration of a motor centre, is the cause of a hyper- or hypo-kinesis. Where the general clinical aspect of a case therefore points to a neurotic origin of the disease, the negative data of morbid anatomy will not disprove such a hypothesis.

With the growing importance of the neurotic element in cutaneous disorders, it was natural that electricity should be called in as a method of treatment, at least wherever the usual therapeutic measures failed to relieve. The number of cases recorded, in which it has been systematically employed, is as yet too small to derive any definite empirical rules of procedure, but sufficient to encourage further attempts in that direction.† Observers will have no difficulty to frame definite indications for each case, and fulfil them by rational applications of the currents after the general remarks at the beginning of this chapter, and the special directions given in the paragraphs on the treatment of nervous disorders.

Applications *in loco morbi* (*i.e.* to the centres and nerve trunk in many cases) and *symptomatis*, of the galvanic current chiefly, and of both poles in succession, will probably be found the most useful in practice, as they are indicated according to the theory.

The central applications will consist chiefly of strong currents to the spinal origin of the nerves, supplying the diseased parts; the peripheral of labile galvanisation of the skin. When vasomotor disturbances are present in the upper part of the body, the sub-aural method may be resorted to (p. 178), and the current localised as much as possible in the cervical roots, by appropriate position of the electrode along the posterior edge of the sterno-mastoid.

Some cases of very obstinate *eczema*, have been reported as having yielded to methodical galvanisation. It would be interesting to test its value in a larger number of similar cases, as well as in lichen, psoriasis, etc.—I have seen good results in neurotic baldness or *alopecia areata*, the growth of the hair following rapidly the first applications; and here, on account of the unpleasant cerebral effects which otherwise ensue, it is necessary to apply the galvanic current by means of two medium sized electrodes, kept near one

---

* "The pathogeny of lead paralysis." *Lancet*, 1880.
As this sheet is going through the press, I am gratified to find that the same view is propounded and developed by Prof. Erb, (*Neurologisches Centralblatt*, No. 21, 1883). The study of peripheral nerve changes (disseminated neuritis, etc.) is being carried out actively. See abstracts in *Brain*, Jan., 1884.

† It is not unlikely, that electricity will prove of use in certain forms of purely local skin affections. Its powerful influence upon the circulation and absorption in (and possibly also on the nutrition of) the diseased tissues, which in the case of the skin are so readily placed under its action, may prove of service in chronic cases, with much infiltration and thickening. Thus labile galvanisation has already been found useful in *acne*. An obvious advantage of electricity as a local excitant, is that it can be applied repeatedly without leaving any traces of its passage (Cf. however, p. 50).

another on the denuded surface. Other forms of loss of hair (after fever and specific disorders) are readily amenable to electricity.

I have also applied galvanism successfully to certain forms of *local asphyxia* with dystrophy of skin and nails (ordinary *chilblains* are usually amenable to either current, locally applied), and in *sclerodermia*. In the latter instance I used strong galvano-faradisation, chiefly to the diseased parts. Certain cases of *hyperidrosis* would seem to offer some chance of success to the electro-therapeutist. *Herpes* which is a typically trophoneurotic disorder scarcely calls for any treatment when occuring in the young. In advanced life, however, it is a more serious disorder, pointing to diminished vitality of the nerve elements, and often accompanied and followed with intense neuralgia. The latter will have to be treated by strong galvanisation or faradisation, on the principles previously stated (p. 172).

APPENDIX.

ELECTROLYSIS.

It is not my intention to give a full account of the application of the galvanic current, to the treatment of various tumours and the like, but merely to indicate the principles, according to which the operations should be carried out.

The utmost confusion and uncertainty reigns in the literature of this subject; and this is chiefly on account of the dense ignorance which so universally prevails, as to the physical laws of electricity, and the consequent neglect of all methods of measurement.[*] And yet in electrotherapeutics, the necessity of such measurements is nowhere more apparent, nor the facility of carrying them out more obvious, a simple voltameter (p. 22) answering the purpose admirably. The hopelessness of arriving at any conclusion, from the recorded observations of electrolysis in *aneurism*, is manifest the moment we look through the tabulated lists of cases given by Poore, Bartholow, and others. Some operators used five cells of a weak (Smee's) battery, others thirty of a strong one (Stöhrer's). Some close the circuit with two needles in the sac, others with one in it and an ordinary electrode outside. Some allow the current to

---

[*] See however Dr. Dixon Mann's excellent paper on "Current Measurement in Electrolysis of the Blood" (*British Medical Journal*, 1878). I must add that Ciniselli, who introduced the electrolytic treatment of aneurism, expressly mentions the fact, that he used currents strong enough to liberate two or three cubic centimetres of the mixed gases, in five minutes; *i.e.* currents of about forty or

flow for twenty minutes, others for a period five times as long. Moreover some used insulated needles, which concentrate the current at the point where it is to act; others non-insulated, which allow perhaps half of it to diffuse through the solid tissues, (which they electrolyse, causing dangerous sloughing).—The matter is further confused by the usual idle talk about the relative merits of batteries made up of "large elements" and of "small elements of low electromotive force," the former being recommended by some, whilst the latter are preferred by others as having a mysterious property, vaguely ascribed to their "tension," of giving less pain. And yet the introduction of a galvanometer in the circuit shows that large cells give a much more powerful current through needles inserted in the tissues, than small ones (see p. 34). It likewise disposes of the argument that the two poles must be inserted into the sac, on account of the resistance of the skin; it is easy to obtain a current of 50 milliamperes (with say 12 to 18 Stöhrers), when a couple of needles are inserted into the sac and the circuit closed with a *very large* plate electrode on the abdomen. Roller electrodes, fig. 45, are not convenient; but in order to avoid the burning sensation and vesication of the skin a layer of modeller's clay may be effectually placed between the plate and skin, as suggested by Dr. Apostoli of Paris.

The principles of the electrolytic action of the current will be found mentioned at page 50. When two needles are placed in a vessel filled with blood, coagulation occurs around the positive (anodic) needle only, the kathode giving rise to a frothy mass without consistence. The weight of the clot is proportional to the current-strength and the duration of its flow. We cannot, however, argue directly from what is observed under experimental conditions, to the process of consolidation of an aneurism. In the one case the blood is altered and stationary; in the other it is living and moves. Moreover, the effect of the current on the sac itself may play an important part in the subsequent extension of the clot; there is no doubt that the effects of an electropuncture are not limited to the time of operation, but are due in no small degree to subsequent deposition of fibrine, around the spots of current-action which act as foreign bodies with reference to the blood.

Unless made of gold or platinum, the positive needles are dissolved in the process of electrolysis. There is no disadvantage, however, in this fact; the oxygen (as well as the chlorine, etc.),

---

fifty milliamperes (see p. 23). The fact that his results were better than those of his followers, who neglected every measurement, is significant.

It is astounding to read how some otherwise able physicians and surgeons, have literally played with electricity in the treatment of aneurisms; their endeavours justify the comparison of the doctor to a man who shuts himself up in a dark room with the patient and the disease, and hits out with a club, in the hope of killing the latter; generally missing it however, but occasionally hitting the former. Among curious therapeutic inanities, on the other hand, I may mention two cases I have recently seen, and in which the galvanic current had been applied by means of sponges to the skin, with a view of coagulating the blood of a thoracic aneurism.

liberated at the needle, combines with it instead of being given off in a gaseous form and the clot is thus more solid. The weight of metal dissolved, is exactly proportional to the quantity of electricity used during the operation.*

From the theoretical considerations just discussed, and the results of my experience, I would advise the operator in a case of aneurism:

1. To introduce into the sac several needles (see fig. 45) connected with the positive pole of the battery, which obviously must have the requisite electromotive force and depolarizing energy, (see page 67) the usual small medical Leclanché elements are scarcely suited to the purpose), and close the circuit with one or more very large plate electrode, which may be applied to any part of the body.

2. To use needles with thoroughly insulated shaft (glass or hard rubber are the best insulators), with spear headed points, taking care that the whole exposed part of the metal be within the sac.

3. To use a measured current strength of 20 or 30 milliamperes at most *per needle*.

4. To allow the current to flow for about half an hour at the first sitting, and increase the duration subsequently according to the results obtained.

The needles should be removed very gently with a rotatory movement, and the external wound immediately closed, with lint soaked in collodion and a bag of ice applied over the aneurism.

The operation being usually almost painless, when properly conducted, no anæsthetic is required, an injection of morphia an hour before may be of use however to subdue the apprehension and agitation of nervous patients.

The operation may be repeated every week, or oftener; the fear of subsequent embolism, though natural, is ungrounded: at least no accident of the kind has hitherto been recorded. With reference to any counter-indications it must be confessed that much of what has been written on the subject is the fruit of theoretical considerations, or vitiated by taking into account operations performed in a senseless manner. Though electrolysis is scarcely to be expected ever to cure an aneurism, at least when aortic, the fact, that it is the only means we possess of relieving the patient, and that it presents no danger when properly carried out, make it desirable that scientifically conducted operations should become frequent enough to give us the data necessary for generalisations of a practical value.

Electrolysis has been used with varying success in the treatment of various solid and cystic tumours, nævi, enlarged glands, ulcers, etc. With reference to *nævi* it may be recommended as a highly efficacious, painless method, and as leaving usually but very faint

---

* It might be convenient for some operators that "Voltametric needles" should be made; *i.e.*, needles of which the uninsulated part has a definite weight of metal, so that its complete destruction in blood should represent a given quantity of electricity. A simple experiment (weighing a needle before and after a current of given strength has been used for a given time) would give the necessary data.

scars. Fine uninsulated needles may be used; when the nævus is very small, one (the kathode) only need be introduced, whilst the circuit is closed by means of a plate electrode in the neighbourhood.—Or two needles may be used, one connected with each pole.

When the tumour is extensive a large number of needles are to be introduced in a circle into its base, and connected alternately with the negative and the positive pole. Care must be taken that the needles of opposite polarity do not come in contact within the nævus, as this would short-circuit the current, and stop all action.

A bichromate battery of 15 cells is usually quite sufficient to electrolyse large nævi. It is best, especially for the small ones, to begin with a few cells only and increase the number gradually until the desired effect is produced. It is impossible to give any definite directions as to the absolute current strength required as this will vary largely from case to case. The operator must depend upon his personal experience for regulating both the intensity and duration of the galvanic action. The object is to solidify the tumour up to a certain point, at which its absorption may take place, but beyond which sloughing takes place. The finger must therefore be educated by practice to discover when that point is reached. In healthy children, however, there is no great harm in making small nævi slough out bodily, as the ulcer that is left, though deep, fills up and skins over very readily, galvanic eschars having the property of not contracting on cicatrisation. Anæsthetics are necessary in the case of children unless the nævus be very small indeed. The pain, however, is quite momentary and no discomfort experienced after the current has been broken—a fact which adds to the superiority of the electrolytic over the other operations for nævus.

In the case of *port wine marks* the object is rather to destroy the skin than cause coagulation. For this purpose numerous fine needles may be inserted superficially, either connected with the negative pole only or alternately with both poles; I have quite lately suggested, but have had no opportunity to test the method, to replace the needles by a small disk of (uncovered) charcoal to be attached to the negative pole and applied to the surface. The new skin is paler than the normal; but as just mentioned no contraction follows the operation. Whilst on this subject I will add that *warts, moles*, and single *hair follicles* may readily be destroyed with the kathodic needle, and superfluous hair and other disfiguring abnormalities of the face removed without much pain and no subsequent scars.

Solid tumours of a malignant nature, namely *cancers*, have been removed by electrolysis. It is useless to attempt to dissolve such growths; the only hopeful plan is to insert numerous large needles along their base so as to separate the whole diseased mass from the healthy underlying tissues and leave the ulcer to granulate and skin over. Powerful currents and long applications are necessary for the purpose. The advantage of the plan is that it avoids the complications attending a knife-operation and its after treatment. Nodules of cancer growing from the cicatrix of previous removal may be conveniently treated by this method. Electrolysis appears to have a sedative effect on the pains of cancer and deserves a more extensive trial in this respect than it hitherto has had.

Fibrous tumours of the uterus are said to have been attacked with the electric negative needle with some success. Their growth seems to be arrested, at least, if their entire removal is not possible.—Various *polypi* and other small superficial tumours have been easily and successfully destroyed by means of the needles and sufficiently strong currents.

Enlarged glands may occasionally require removal, and electrolysis being an easy and painless method should be first tried. The two poles should be introduced and 6-10 cells used. (External faradisation with strong currents sometimes acts beneficially on lymphatic enlargements).

In electrolysing *cystic tumours* our aim should be less to destroy the growth, than so to alter its contents and the internal lining as to excite absorption. Successful cases of hydatids of the liver, cystic goitre, etc. have been reported. In some cases a solution of iodide of potassium has been injected into the cyst (*e.g.* in hydrocele), by means of a trocar through which a long insulated needle was then passed, connected with the negative pole. The anode being placed on the surface. Nascent iodine in very small quantities is supposed to be formed at the surface of the lining membrane and so assist the effect of the electrolysis.° A current of 10 or 15 cells for about twenty minutes seems to be sufficient to obtain permanent results in hydrocele.—In the case of *sebaceous* tumours the following method may be tried where the simpler plan of removal by excision offers any objections. A small bulbar electrode, with insulated stem and connected with the negative pole is introduced into the tumour. The positive electrode, in the shape of a plate (p. 89) with a hole in its centre through which the tumour protrudes, is applied outside, and the current made as strong as bearable. The kathodic bulb is then made to act upon every point of the lining membrane so as to destroy its vitality.

Strictures of the urethra have frequently been removed by electrolysis. A bougie a size larger than the stricture with a metal tip connected with the negative pole is introduced, and made to press against the stricture. A current of 6 gradually increased to 12 or 15 cells is then made by applying a very large electrode to the thigh or abdomen, and the bougie gradually slips through the electrolysed tissues. The operation is repeated at intervals of a few days if necessary. The advantages of the method are the absence of danger and fever, as well as of consecutive contraction of the strictured part. Stricture of the *œsophagus* might be attacked on the same principle.

Indolent or painful *ulcers* may be much improved by slight electrolysis. This may be effected by means of a silver and a zinc plate connected with a piece of telegraph wire and applied for

---

* It must be noted, however, that electrolysis has succeeded in cases of goitre where previous treatment by means of iodine injections had failed.

twenty-four hours or more, the silver being fixed to the ulcer, the zinc to the skin with a piece of moistened linen intervening; or a stronger current may be applied daily by means of metal electrodes for a shorter period. Apostoli recommends the following method:—The skin around the ulcer is painted over with collodion; the ulcer filled with modeller's clay and the metal electrode made to cover the latter.—In all applications of long duration measures must be adopted against the dessication of the electrodes.

The anode is said to diminish, the kathode to increase, the suppuration of the ulcer. It must be noted that when the simple silver-zinc couple is applied as above described the external circuit is formed by the wire, the liquids of the body represent the exciting fluid of the cell, the current flowing inside the body from the zinc to the silver, and the effect is the same therefore as when the kathode of the battery is applied to the ulcer.

INDEX.

Abdomen, 190
Abdominal neuralgia, 192
Absolute galvanometers, 28
ACC (anodic closure contraction), 100
Accessories, 82
Acne, 197
Acoustic nerve, 167, 188
ADC (anodic duration contraction), 100
After effects, 112
Alopecia, 197
Alternatives (voltaic), 108
Amalgamation, 55
Amenorrhœa, 195
Ampere (or Weber), unit of current strength, 15
Anæmia, cerebral, 177
Anæsthesia, 171
Anelectrotonus, 101
Aneurism, 198
Angina pectoris, 190
Anode, 5
Anodic reactions, 101, 105
AOC (anodic closure contraction), 100
Aphasia, 176
Aphonia, 185
Arm, nerve of, 186
Arthritis, 194
Articular rheumatism, 193
Arrangement of elements in batteries, 32
Ascites, 192
Asphyxia, 185
   ,, of extremities, 198
Asthma, 190
Ataxy, locomotor (see tabes dorsalis)
Atonic (conditions of digestive organs), 191
Atrophy, simple and degenerative, 133

Bath (electric), 102
Batteries, 51
Bichromate of potash element, 69
Bladder, 194
Blepharospasm, 170
Brain, electrisation of, 118, 174
Break (see opening)
Brush (faradic), 91
Bulbar paralysis, 178
Bunsen element, 65

Cancer, electrolysis of, 201
Catalysis, 147
Cathode (see kathode)
Cells (see elements)
Central galvanisation, 162
Central treatment, 166, 174
Cerebral diseases, 123
   ,, hæmorrhage and softening, 176
   ,, effects of galvanism, 118
Cervico-brachial and occipital neuralgiæ, 174
Chemical effects, 49 (see electrolysis)
Chilblains, 198
Choice of current (for treatment), 150
   ,, ,, electrodes, 153
Chorea, 181
Chloride of silver element, 59
Cirrhosis (see degenerative atrophy)
Clay electrodes, 199, 202
Closure (make), 98
Collectors, 84
Commutator, 94
Comparison of excitations, 135
Conditions of equal excitations, 138
Conductivity (see resistance)
Connections, 89
Constancy of current, 38
Constant current (see galvanism)
Contractures, 168
Cramp (see spasm)
Current, 3
   ,, Choice of, 150
Current alternator, commutator and combiner, 95
   ,, measurement, 22
   ,, reverser (see commutator)
Currents (weak, strong etc,) used in medical practice, 15, 151
Cystic tumours, 202

Daniell's cell, 62
Degeneration (degenerative atrophy), 127
Density, 41
Derived currents, 40
Diabetes, 182
Dial-collector, 86
Diaphragm (see phrenic nerve)

INDEX.

Diffusion of current, 42
Diphtheritic paralysis, 187
Direction of current, 104, 107, 147
Disposition of electrodes, 153
Dyspepsia, 191
Dynamic electricity, 1

Ear (diseases of the), 188
Ear (electrophysiology), 117
Eczema, 197
Electricity definition, 1
Electrification, 1
Electrisation (therapeutical), 156
Electrodes, 89, 153
   ,,   large, medium, small, etc., 90
Electrodiagnosis, 120
   ,,   (practice of), 143
Electrolysis, 50
   ,,   of tumours, 198
Electromotive force, 5, 18
Electromuscular and nervous excitability, 123, 125
Electrophysiology, 97
Electroprognosis, 135
Electrotherapeutics, 146
Electrotonus, 101, 110
Elements (galvanic), 51
Enteralgia, 192
Enuresis, nocturna, 194
Epidermis, resistance of, 44
Epilepsy, 181
Excitability of nerve and muscle, 97
   ,,   faradic, 109
   ,,   galvanic, 97
   ,,   ,,   in disease, 122
Exophthalmic goitre, 182
Extra-current (primary), 71
Eye reactions to galvanism, 116
   ,,   Diseases of, 188

Facial neuralgia, 173
   ,,   paralysis, 133, 184
   ,,   spasm, 170
Fallacies of electrodiagnosis, 142
Faradisation, 160
   ,,   general, 162
Faradism, 2, 69
Faradic apparatus, 76
Faradic excitation of motor nerves, 109

Galvanisation, 158
Galvanic cell, 9
Galvanism or Voltaism, 1
Galvanic apparatus, 51
Galvanic excitation of motor nerves, 97
Galvano-faradisation, 163
Galvanometer, 25
   ,,   in diagnosis, 140
Galvano-puncture, 174
General electrisation, 162

Glands, enlarged, 202
Graduation of current, 82
Graphical representation of electric phenomena, 16
Grenet element, 67
Grove element, 65

Hair, removal by electrolysis, 201
Headache, 178
Heart, excitation of, 114
   ,,   neuroses of, 190
Hemicrania, 183
Hemiplegia, 176
Herpes, 198
Histrionic spasm, 170
Human body as conductor, 43
Hydrocele, electrolysis in, 202
Hyperæsthesia, 172
Hyperæmia (cerebral), 177
Hyperidrosis, 198
Hypochondriasis, 181
Hysteria, 181

Impotence, 195
Incontinence of urine, 194
Induced currents (see faradism)
Interrupted galvanisation, 160
Intercostal neuralgia, 174
Intestinal obstruction, 192

Katelectrotonus, 101
Kathode, 5
Kathodic reactions, 105
KCC (kathodic closure contraction), 99
KDC (kathodic duration contraction), 100
KOC (kathodic opening contraction), 100

Labile galvanisation, 159
Labio-glosso-pharyngeal paralysis (bulbar), 178
Larynx, 185
Late rigidity, 176
Lateral sclerosis, 179
Law of contractions, 100
   ,,   of sensory reactions, 115
Lumbago (see Rheumatism)
Lungs (neuroses of), 190
Lead palsy, 187
Leclanché element, 56
Liver (congestion), 192

Magneto-electric instruments, 80
Magnetism (terrestrial), 30
Malingering, 123
Make (see closure)
Maximal and minimal contraction, 122
Measurements, 18, 20, 22, 28

INDEX.

Meningitis (spinal), 179
Menorrhagia, 195
Meteorism, 192
Methods of electrisation (polar), 147
Metritis, 196
Mechanical effects (osmosis), 47
Megrim, 183
Measurements, 18
Milk (secretion of) 196
Milliampere (formerly called Milliweber), 13, 28
Modal alterations of contraction, 122
Modifying influence (electrotonus), 146
Monoplegia, 177
Motor nerves, 97
  ,, electrophysiology, 97
  ,, electrodiagnosis, 125
Motor points, 114 (cf. Plates).
Muscle, excitation of, 113, 145
Muscular atrophy, 180, 193
  ,, rheumatism, 193
Myelitis, 178

Nævi, electrolysis of, 200
  Negative pole (see kathode)
Nerves (electro-physiology), 97
Neuralgia (and neuralgiform affections), 172
Neurasthenia, 181
Neuritis, 184
Neuroses, 181, 190
Normal polar formula (see law of contractions).

Obstruction, 192
  Ohm (unit of resistance) 13
Ohm's law, 12
Opening (break) of current, 98
Optic nerve, electro-physiology, 116

Pain, 172
  Painful pressure points, 169
Paralysis, 165
  ,, agitans, 182
  ,, infantile, 179
Paraplegia, 179
Paresis, 165
Partial RD, 131
Peripheral nerve lesions, 125, 184
  ,, treatment, 166
Peripolar zone, 103
Peroxide of manganese element, 56
Phrenic nerve, 185
Plexus, brachial and lumbar, 134
Plug collector, 84
Physical effects, 48
Polar effects, 104
  ,, method of treatment, 147
  ,, zone, 103

Polarisation of batteries, 38
  ,, cells, 68
  ,, of nerves (see electrotonus)
  ,, of tissues, 50, 119
Pole (positive or anode, negative or kathode), 5
Poliomyelitis, 179
Port wine marks, electrolysis of, 201
Position of electrodes, 153
Posology, 151
Potentials, 2, 42
Ptosis (see eye)
Pressure points, 170, 172
Primary current, 75
Progressive muscular atrophy, 180

Qualitative and quantitative alterations, 122

Railway spine, 180
  Reaction of degeneration (RD), 127
Rectum reaction, normal, 123
Reflex action in electrisation, 167
Regeneration, 127
Resistance, 10, 20
Respiratory spasms, 170
  ,, neuroses, 190
Reversed galvanisation, 160
Rheophores, 88
Rheostat, 20, 82
Rheumatism, 193
Roots, cervical and lumbar, 134

Sciatica, 174
  Sclerodermia, 198
Scrivener's palsy, 182
Secondary coil, 72
Secondary current, 75
Secretory nerves, 119
Sensory nerves, (electrophysiology of), 115
Serial alterations, 122
Sick headache, 183
Skin (resistance of), 45
  ,, faradisation, 160
  ,, diseases, 196
Sledge collector, 85
Smee element, 66
Spasms, 168
Spinal cord, 118
  ,, diseases, 123
Spinal irritation, 180
Spleen, 192
Stabile galvanisation, 158
Static electricity, 1
Stöhrer's elements, 66
Strength (current) 14, 20
Strictures, electrolysis of, 202
Strong (medical) currents, 15
Subaural galvanisation, 164

INDEX.

Sulphate of copper element, 62
  „ „ mercury „   64
Sympathetic, 118, 164
Symptomatic treatment, 166
Syphilitic paralysis, 187

Tabes dorsalis, 179
  Tangent galvanometer, 26
Taste (galvanic), 117
Tension, 34
Tetanus, 182
  „ (duration contraction), 100
Tic convulsif (see facial spasm)
  „ douloureux (see neuralgia)
Tinnitus aurium, 189
Torticollis, 170
Toxic paralysis, 187
Traumatic paralysis, 127, 184
Treatment " in loco morbi " and " in loco symptomatis ", 165
Trigeminus, 173
Trismus, 170
Trophic nerves, 119, 131
  „ changes, 127, 196
Tympanitis, 192

Ulcers, electrolysis, 202
  Units, 12
Urinary organs, 194
Uterine disorders, 195
Unipolar method, 147

Vasomotor nerves, 118
  Virtual electrode, 104
Volt (unit of electromotive force), 13
Voltaic alternatives, 108
Voltameter, 23
Vomiting, 191

Warts, electrolysis of, 201
  Weber (see Ampere),
Writers' cramp, 182
Wryneck, 170

Zinc carbon element, 66

MOTOR POINTS AND MOTOR ROOTS.

THE accompanying plates illustrate the position of the chief nerve trunks and motor points of muscles (see page 113) accessible to localised electrical excitation. The student should diligently practice upon himself the art of applying the current to individual muscles and nerves, he will thereby gain a much more rapid and perfect understanding of the process, which when properly conducted is by no means painful.

The facility with which the excitation can be localised varies much in different individuals, especially with regard to the deeper structures. The greater the accumulation of subcutaneous fat the more difficult it is to obtain contractions of single muscles. The strength of current required, and the degree of pressure to be exercised upon the electrode vary a good deal in each case according to the distance and nature of the tissues, intervening between the surface and the particular structure to be reached.

I have added short tables showing the nervous supply of muscles, and the relations of nerve trunks to the cerebro-spinal motor roots (cf. p. 134). Some nerves (*e.g.* ulnar) are thereby shown to arise from circumscribed portions of the spinal grey matter, whilst others (*e.g.* musculospiral) receive fibres, from a much more extensive area. Physiological experiments and clinical observation, however, alone can determine in the latter case with how many roots each muscle supplied by such a nerve is in relation.

As far as our present knowledge goes, we may, I think, assume as a law, *that the greater the number of co-ordinated movements in which a muscle takes part, either as an associate or an antagonist, the greater the number of spinal centres (or spinal segments) with which it is connected.* It is probable that every group of muscles producing flexion or extension, or adduction, etc., of an articulation is under the dependence of a chief centre by the acting of which its simple, forcible, contraction is brought about. In addition to this we must assume subsidiary centres, or at least subsidiary functions of the chief centres, through which the necessary simultaneous concourse of other muscles is ensured. Thus we find the 8th cervical and 1st dorsal roots, which are chiefly distributed to the flexors of the hand and fingers, send a branch to the musculospiral which innervates the extensors; now it is easy to demonstrate that a firm closure of the hand is impossible unless the wrist is supported by a contraction of the extensors. On the contrary the intrinsic muscles of the hand, the co-operation of which with the long extensors is not obligatory, receive their almost exclusive supply through the ulnar nerve, from the lower two roots of the cervical plexus.*

This question of spinal motor centres is peculiarly interesting to the electro-therapeutist who by skilful diagnostic applications of the currents, combined with a knowledge of the anatomy and physiology of the neuromuscular apparatus, possesses the means of throwing much light upon a number of morbid processes of which atrophic paralysis of muscles is a prominent symptom.

---

* Prof. Erb has lately discovered a motor point in the neck, (see Plate I.), corresponding to the united 5th and 6th cervical roots, excitation at which throws into contractions the deltoid and some scapular muscles, together with the *biceps*, *brachialis* and *supinator longus*. The latter three are flexors of the arm; here then we have a good illustration of the common central innervation of muscles physiologically related to one another, though supplied by different nerve trunks. (See also, "On the Functions of the Brachial Plexus," Dr. Poore's Bradshaw Lecture, *Lancet*, 1881, vol. ii.)

BULBAR ROOTS.

III. OCULOMOTOR.
    Levator palpebræ.
    Rectus superior.
    Rectus inferior.
    Rectus internus.
    Obliquus inferior.
        Iris (constrictor fibres).
        Ciliary muscle.

IV. TROCHLEAR.
    Obliquus superior.

VI. ABDUCENS.
    Rectus externus.

V. TRIGEMINUS.
    Inferior maxillary division.
      Temporalis.
      Masseter.
      Pterygoids.
    Inferior dental nerve.
      Mylohyoid.
      Digastric (anterior portion).

VII. FACIAL.
1. *Through the spheno-palatine ganglion.*
    Azygos uvulæ.
    Levator palati.
2. *Through the otic ganglion.*
    Tensor tympani.
    Tensor palati.
3. *Laxator tympani.*
    Stapedius.
4. *Chorda tympani.*
    Lingualis.
5. *Posterior auricular.*
    Retrahens aurem.
    Occipito-frontalis.
6. Stylohyoid.
    Digastric (posterior portion).
    Platysma myoides (with 3rd cervical).
7. Muscles of the face.
    Muscles of the external ear.

XI. SPINAL ACCESSORY.
    Stylopharyngeus.
    Constrictors of pharynx.
    Laryngeal muscles.
    Sternomastoid (with 2nd cervical).
    Trapezius (with 3rd cervical).

XII. HYPOGLOSSUS.
    Linguales.
    Styloglossus.
    Hyoglossus.
    Genio-hyoglossus.
    Omohyoid   ⎫
    Sternohyoid   ⎬ (with 2nd and 3rd cervical)
    Sternothyroid ⎭

CERVICAL AND FIRST DORSAL ROOTS.

I. Deep rotators, &c., of head.

II. Complexus.
Splenius.
Trachelo-mastoid.

III. Platysma (with *facial*).
Levator anguli scapulæ.
Trapezius.

IV. V. PHRENIC (diaphragm).
Scaleni.

(IV.) V. DORSALIS SCAPULÆ.
Rhomboids.
SUBCLAVIUS.

V. VI. POSTERIOR THORACIC.
Serratus magnus.
SUPRASCAPULAR.
Supra- and infra-spinatus.

V. VI. VII. EXTERNAL ANTERIOR THORACIC.
Pectoralis major (with *int. ant. thor.*)
MUSCULOCUTANEOUS.
Coraco-brachialis.
Biceps.
Brachialis anticus (with *musculospiral*).

V. VI. VII.
(VIII. I DORS.) } SUBSCAPULARIS.
Subscapularis.
Teres major.
Latissimus dorsi.
CIRCUMFLEX.
Deltoid.
Teres minor.
MUSCULOSPIRAL.
Triceps.
Anconeus.
Supinators.
Long extensors of hand and fingers.

(V. VI. VII.)
VIII. I DORS. } MEDIAN.
Pronators.
Long flexors of hand and fingers (with *ulnar*).
Lumbricales (1st and 2nd).
Opponens and abductor pollicis.
Flexor brevis pollicis (outer half).

VIII. I DORS. INTERNAL ANT. THORACIC.
Pectorales.

ULNAR.
Flexor carpi ulnaris.
Flexor profundus digitorum (with *median*).
Interossei.
Lumbricales 3rd and 4th.
Adductor and flexor brevis pollicis (inner half).
Abductor, opponens and flexor brev. min. digiti.
Palmaris brevis.

LUMBAR AND SACRAL ROOTS.

I. Rectus abdominis.
Obliquus internus.

I. II. **Cremaster.**

II. III. Psoas, iliacus.

II. III. IV. ANTERIOR CRURAL.
Sartorius.
Pectineus (with *obturator*).
Rectus femoris.
Vastus externus et internus.
Crureus.

OBTURATOR.
Obturator externus.
Pectineus.
Gracilis.
Adductor **magnus** (with *sciatic*).
Adductor longus.
Adductor brevis.

IV. V. SUPERIOR GLUTEAL.
Gluteus medius.
Gluteus minimus.
Tensor vaginæ femoris.

IV. V. I. II. III. Obturator internus.
Gemelli.
Quadratus femoris.

GREAT SCIATIC.
Flexors of **leg**.
Adductor magnus **(with** *obturator*).

INTERNAL POPLITEAL.
Gastrocnemius, and other flexors of ankle
Flexors of toes.
Intrinsic muscles of foot.

EXTERNAL POPLITEAL, OR PERONEAL.
Tibialis anticus.
Extensor of toes.
Peronei muscles.

II. Pyriformis.

III. IV. PUDIC NERVE.
Perineal muscles.
Levator ani.

IV. Sphincter ani.
Coccygeus.

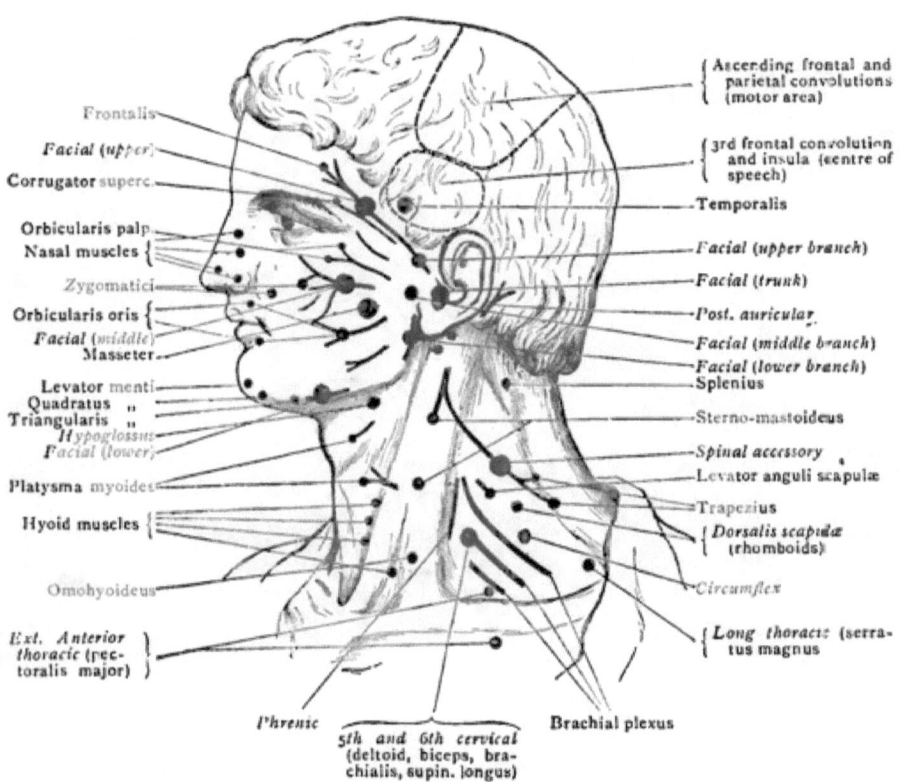

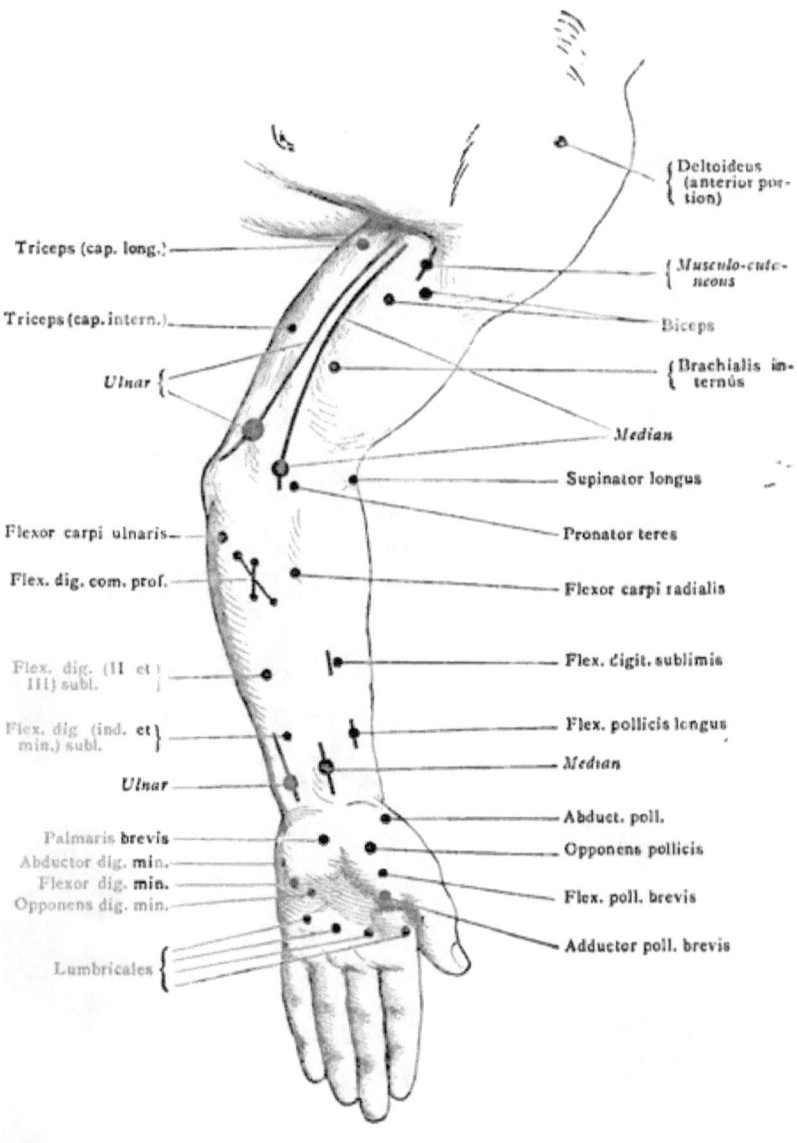

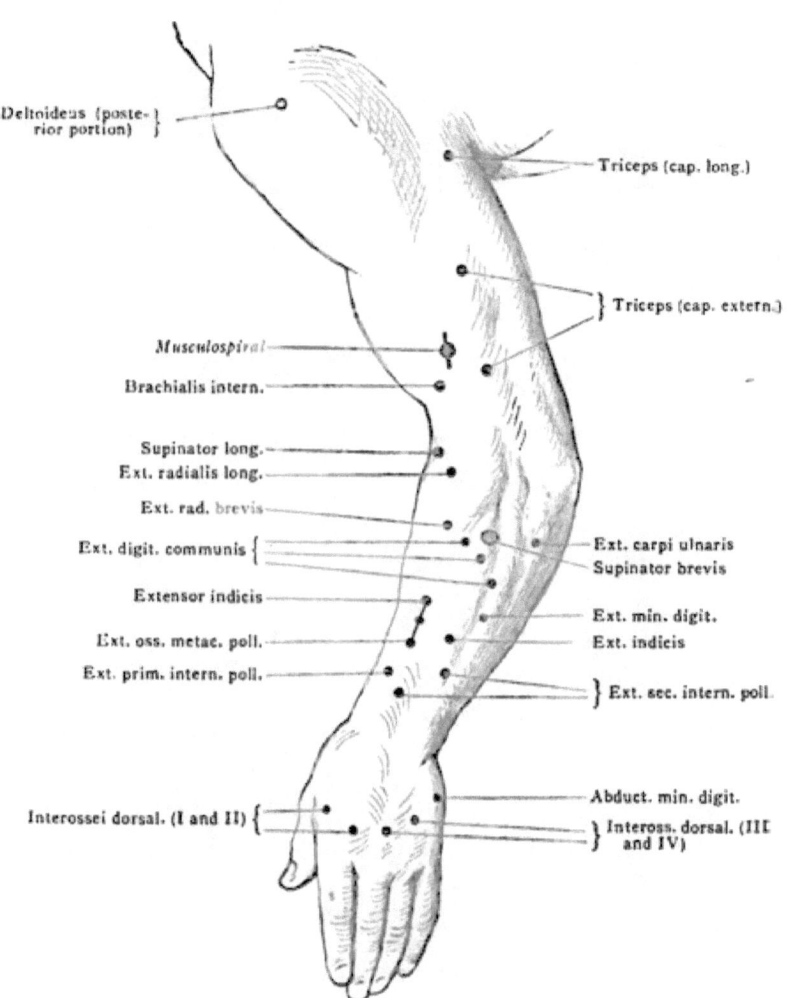

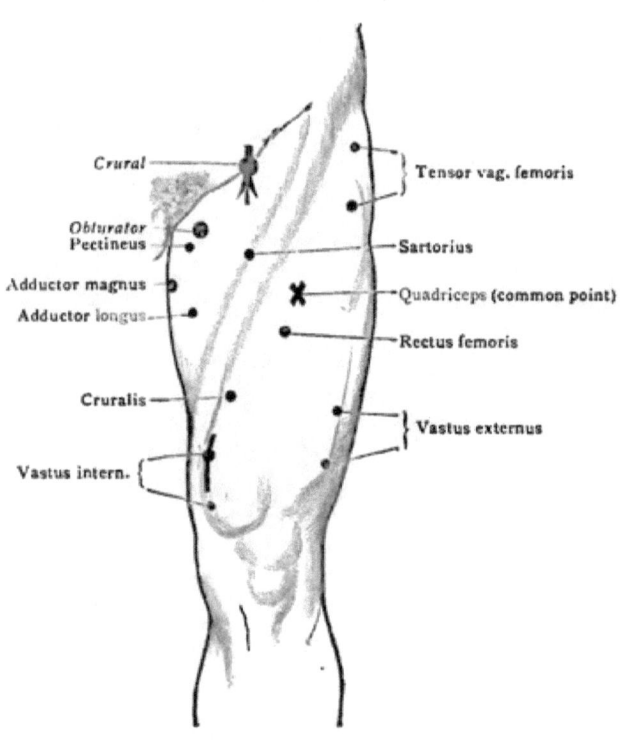

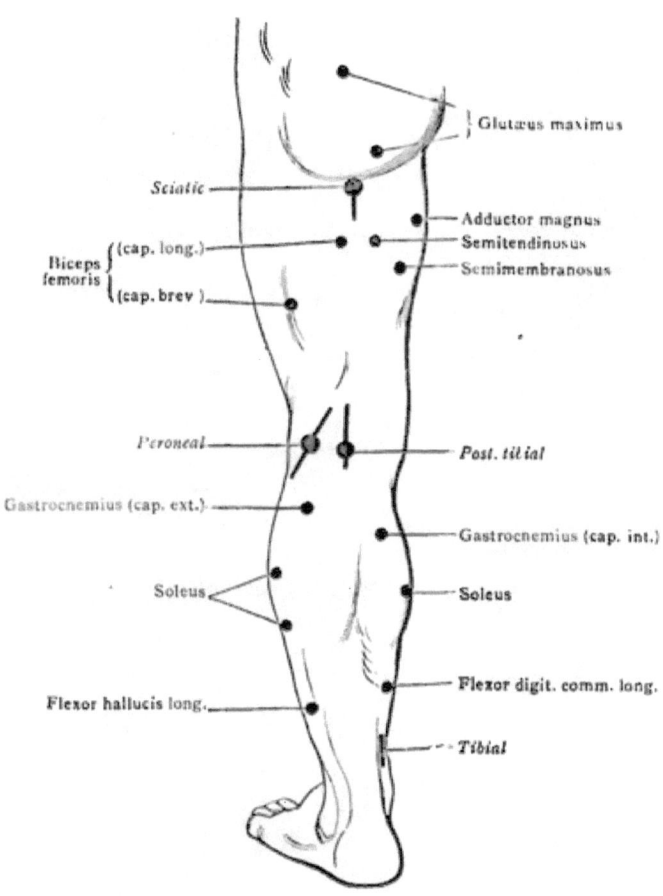

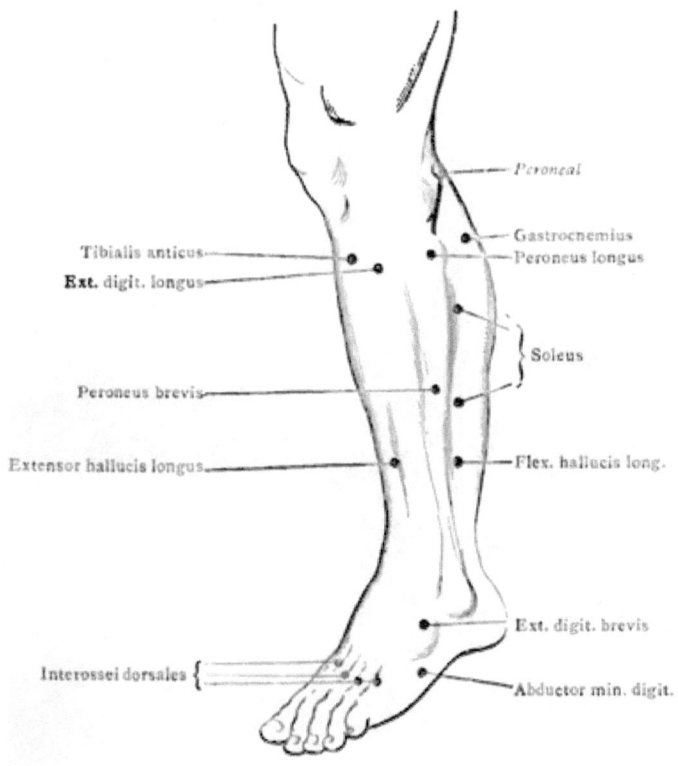

November, 1884.

CATALOGUE OF WORKS
PUBLISHED BY
H. K. LEWIS
136 GOWER STREET, LONDON, W.C.

G. GRANVILLE BANTOCK, M.D., F.R.C.S. EDIN.
Surgeon to the Samaritan Free Hospital for Women and Children.

I.

ON THE USE AND ABUSE OF PESSARIES. Second Edition, with Illustrations, 8vo, 5s.

II.

A PLEA FOR EARLY OVARIOTOMY. Demy 8vo, 2s.

FANCOURT BARNES, M.D., M.R.C.P.
Physician to the Chelsea Hospital for Women; Obstetric Physician to the Great Northern Hospital, &c.

A GERMAN-ENGLISH DICTIONARY OF WORDS AND TERMS USED IN MEDICINE AND ITS COGNATE SCIENCES. Square 12mo, Roxburgh binding, 9s.

ASHLEY W. BARRETT, M.B. LOND., M.R.C.S., L.D.S.
Dental Surgeon to the London Hospital, &c.

DENTAL SURGERY FOR GENERAL PRACTITIONERS AND STUDENTS OF MEDICINE. With Illustrations, crown 8vo.
[*In the press.*

Lewis's Practical Series].

ROBERTS BARTHOLOW, M.A., M.D., LL.D.
Professor of Materia Medica and Therapeutics, in the Jefferson Medical College of Philadelphia, &c., &c.

I.

A TREATISE ON THE PRACTICE OF MEDICINE, FOR THE USE OF STUDENTS AND PRACTITIONERS. Fifth Edition, with Illustrations, large 8vo, 21s. [*Just published.*

II.

A PRACTICAL TREATISE ON MATERIA MEDICA AND THERAPEUTICS. Fifth Edition, Revised and Enlarged, 8vo, 18s.
[*Just published.*

GEO. M. BEARD, A.M., M.D.
Fellow of the New York Academy of Medicine; Member of the American Academy of Medicine, &c.
AND
A. D. ROCKWELL, A.M., M.D.
Fellow of the New York Academy of Medicine; Member of the American Academy of Medicine, &c.

A PRACTICAL TREATISE ON THE MEDICAL AND SURGICAL USES OF ELECTRICITY. Including Localized and General Faradization; Localized and Central Galvanization; Franklinization; Electrolysis and Galvano-Cautery. Fourth Edition. With nearly 200 Illustrations, roy. 8vo, 28s. [*Just published.*

A. HUGHES BENNETT, M.D.
Member of the Royal College of Physicians of London; Physician to the Hospital for Epilepsy and Paralysis, Regent's Park, and Assistant Physician to the Westminster Hospital.

I.
A PRACTICAL TREATISE ON ELECTRO-DIAGNOSIS IN DISEASES OF THE NERVOUS-SYSTEM. With Illustrations, 8vo, 8s. 6d.

II.
ILLUSTRATIONS OF THE SUPERFICIAL NERVES AND MUSCLES, WITH THEIR MOTOR POINTS, A knowledge of which is essential in the Art of Electro-Diagnosis. (Extracted from the above). 8vo, paper cover 1s. 6d., cloth 2s.

III.
ON EPILEPSY: ITS NATURE AND TREATMENT. 8vo, 2s. 6d.

DR. THEODOR BILLROTH.
Professor of Surgery in Vienna.

GENERAL SURGICAL PATHOLOGY AND THERAPEUTICS. In Fifty-one Lectures. A Text-book for Students and Physicians. With additions by Dr. ALEXANDER VON WINIWARTER, Professor of Surgery in Luttich. Translated from the Fourth German edition with the special permission of the Author, and revised from the Tenth edition, by C. E. HACKLEY, A.M., M.D. Copiously illustrated, 8vo, 18s.

G. H. BRANDT, M.D.

I.
ROYAT (LES BAINS) IN AUVERGNE, ITS MINERAL WATERS AND CLIMATE. With Frontispiece and Map. Second edition, crown 8vo, 2s. 6d.

II.
HAMMAM R'IRHA, ALGIERS. A Winter Health Resort and Mineral Water Cure Combined. With Frontispiece and Map, crown 8vo, 2s. 6d.

GURDON BUCK, M.D.

CONTRIBUTIONS TO REPARATIVE SURGERY; SHOWing its Application to the Treatment of Deformities, produced by Destructive Disease or Injury; Congenital Defects from Arrest or Excess of Development; and Cicatricial Contractions from Burns. Illustrated by numerous Engravings, large 8vo, 9s.

ALFRED H. CARTER, M.D. LOND.
Member of the Royal College of Physicians; Physician to the Queen's Hospital, Birmingham; Examiner in Medicine for the University of Aberdeen, &c.

ELEMENTS OF PRACTICAL MEDICINE. Third Edition, crown 8vo, 9s. [*Now ready.*]

P. CAZEAUX.
Adjunct Professor in the Faculty of Medicine of Paris, &c.
AND
S. TARNIER.
Professor of Obstetrics and Diseases of Women and Children in the Faculty of Medicine of Paris.

OBSTETRICS: THE THEORY AND PRACTICE; including the Diseases of Pregnancy and Parturition, Obstetrical Operations, &c. Seventh Edition, edited and revised by ROBERT J. HESS, M.D., with twelve full-page plates, five being coloured, and 165 wood-engravings, 1081 pages, roy. 8vo, 35s. [*Just ready.*]

JOHN COCKLE, M.A., M.D.
Physician to the Royal Free Hospital.

ON INTRA-THORACIC CANCER. 8vo, 4s. 6d.

W. H. CORFIELD, M.A., M.D. OXON.
Professor of Hygiene and Public Health in University College, London.

DWELLING HOUSES: Their Sanitary Construction and Arrangements. Second Edition, with Illustrations. [*In the press.*

J. THOMPSON DICKSON, M.A., M.B. CANTAB.
Late Lecturer on Mental Diseases at Guy's Hospital.

THE SCIENCE AND PRACTICE OF MEDICINE IN RELATION TO MIND, the Pathology of the Nerve Centres, and the Jurisprudence of Insanity, being a course of Lectures delivered at Guy's Hospital. Illustrated by Chromo-lithographic Drawings and Physiological Portraits. 8vo, 14s.

HORACE DOBELL, M.D.
Consulting Physician to the Royal Hospital for Diseases of the Chest, &c.

I.
ON DIET AND REGIMEN IN SICKNESS AND HEALTH, and on the Interdependence and Prevention of Diseases and the Diminution of their Fatality. Seventh Edition, 8vo, 10s. 6d.

II.
AFFECTIONS OF THE HEART AND IN ITS NEIGHBOURHOOD. Cases, Aphorisms, and Commentaries. Illustrated by the heliotype process. 8vo, 6s 6d.

JOHN EAGLE.
Member of the Pharmaceutical Society.

A NOTE-BOOK OF SOLUBILITIES. Arranged chiefly for the use of Prescribers and Dispensers. 12mo, 2s. 6d.

JOHN ERIC ERICHSEN.
Holme Professor of Clinical Surgery in University College; Senior Surgeon to University College Hospital, &c.

MODERN SURGERY; Its Progress and Tendencies. Being the Introductory Address delivered at University College at the opening of the Session 1873-74. Demy 8vo, 1s.

DR. FERBER.

MODEL DIAGRAM OF THE ORGANS IN THE THORAX AND UPPER PART OF THE ABDOMEN. With Letter-press Description. In 4to, coloured, 5s.

AUSTIN FLINT, JR., M.D.
Professor of Physiology and Physiological Anatomy in the Bellevue Medical College, New York; attending Physician to the Bellevue Hospital, &c.

I.

A TEXT-BOOK OF HUMAN PHYSIOLOGY; DESIGNED for the Use of Practitioners and Students of Medicine. New edition, Illustrated by plates, and 313 wood engravings, large 8vo, 28s.

II.

THE PHYSIOLOGY OF THE SPECIAL SENSES AND GENERATION; (Being Vol. V of the Physiology of Man). Roy. 8vo, 18s.

J. MILNER FOTHERGILL, M.D.
Member of the Royal College of Physicians of London; Physician to the City of London Hospital for Diseases of the Chest, Victoria Park, &c.

I.

THE HEART AND ITS DISEASES, WITH THEIR TREATMENT; INCLUDING THE GOUTY HEART. Second Edition, entirely re-written, copiously illustrated with woodcuts and lithographic plates. 8vo. 16s.

II.

INDIGESTION, BILIOUSNESS, AND GOUT IN ITS PROTEAN ASPECTS.

PART I.—INDIGESTION AND BILIOUSNESS. Post 8vo, 7s. 6d.
PART II.—GOUT IN ITS PROTEAN ASPECTS. Post 8vo, 7s. 6d.

III.

HEART STARVATION. (Reprinted from the Edinburgh Medical Journal), 8vo, 1s.

ERNEST FRANCIS, F.C.S.
Demonstrator of Practical Chemistry, Charing Cross Hospital.

PRACTICAL EXAMPLES IN QUANTITATIVE ANALYSIS, forming a Concise Guide to the Analysis of Water, &c. Illustrated, fcap. 8vo, 2s. 6d.

HENEAGE GIBBES, M.D.
Lecturer on Physiology and Histology in the Medical School of Westminster Hospital; late Curator of the Anatomical Museum at King's College.

PRACTICAL HISTOLOGY AND PATHOLOGY. Second Edit. revised and enlarged. Crown 8vo, 5s.

C. A. GORDON, M.D., C.B.
Deputy Inspector General of Hospitals, Army Medical Department.

REMARKS ON ARMY SURGEONS AND THEIR WORKS. Demy 8vo, 5s.

W. R. GOWERS, M.D., F.R.C.P. M.R.C.S.
Physician to University College Hospital, &c.

DIAGRAMS FOR THE RECORD OF PHYSICAL SIGNS. In books of 12 sets of figures, 1s. Ditto, unbound, 1s.

SAMUEL D. GROSS, M.D., LL.D., D.C.L., OXON.
Professor of Surgery in the Jefferson Medical College of Philadelphia.

A PRACTICAL TREATISE ON THE DISEASES, INJURIES, AND MALFORMATIONS OF THE URINARY BLADDER, THE PROSTATE GLAND; AND THE URETHRA. Third Edition, revised and edited by S. W. GROSS, A.M., M.D., Surgeon to the Philadelphia Hospital. Illustrated by 170 engravings, 8vo, 18s.

SAMUEL W. GROSS, A.M., M.D.
Surgeon to, and Lecturer on Clinical Surgery in, the Jefferson Medical College Hospital, and the Philadelphia Hospital, &c.

A PRACTICAL TREATISE ON TUMOURS OF THE MAMMARY GLAND: embracing their Histology, Pathology, Diagnosis, and Treatment. With Illustrations, 8vo, 10s. 6d.

WILLIAM A. HAMMOND, M.D.
Professor of Mental and Nervous Diseases in the Medical Department of the University of the City of New York, &c.

I.
A TREATISE ON THE DISEASES OF THE NERVOUS SYSTEM. Seventh edition, with 112 Illustrations, large 8vo, 25s.

II.
A TREATISE ON INSANITY. Large 8vo, 25s.
[*Just published.*

III.
SPIRITUALISM AND ALLIED CAUSES AND CONDITIONS OF NERVOUS DERANGEMENT. With Illustrations, post 8vo, 8s. 6d.

ALEXANDER HARVEY, M.A., M.D.
Emeritus Professor of Materia Medica in the University of Aberdeen; Consulting Physician to the Aberdeen Royal Infirmary, &c.

FIRST LINES OF THERAPEUTICS; as based on the Modes and the Processes of Healing, as occurring Spontaneously in Disease; and on the Modes and the Processes of Dying, as resulting Naturally from Disease. In a series of Lectures. Post 8vo, 5s.

ALEXANDER HARVEY, M.D.
Emeritus Professor of Materia Medica in the University of Aberdeen, &c.
AND
ALEXANDER DYCE DAVIDSON, M.D.
Professor of Materia Medica in the University of Aberdeen.

SYLLABUS OF MATERIA MEDICA FOR THE USE OF TEACHERS AND STUDENTS. Based on a selection or definition of subjects in teaching and examining; and also on an estimate of the relative values of articles and preparations in the British Pharmacopœia with doses affixed. Seventh Edition, 32mo.
[*In preparation.*

GRAILY HEWITT, M.D.
Professor of Midwifery and Diseases of Women in University College, Obstetrical Physician to University College Hospital, &c.

OUTLINES OF PICTORIAL DIAGNOSIS OF DISEASES OF WOMEN. Fol. 6s.

BERKELEY HILL, M.B. LOND., F.R.C.S.
Professor of Clinical Surgery in University College; Surgeon to University College Hospital and to the Lock Hospital.

THE ESSENTIALS OF BANDAGING. For Managing Fractures and Dislocations; for administering Ether and Chloroform; and for using other Surgical Apparatus. Fifth Edition, revised and much enlarged, with Illustrations, fcap. 8vo, 5s.

BERKELEY HILL, M.B. LOND., F.R.C.S.
Professor of Clinical Surgery in University College; Surgeon to University College Hospital and to the Lock Hospital.

AND

ARTHUR COOPER, L.R.C.P., M.R.C.S.
Late House Surgeon to the Lock Hospital, &c.

I.

SYPHILIS AND LOCAL CONTAGIOUS DISORDERS. Second Edition, entirely re-written, royal 8vo, 18s.

II.

THE STUDENT'S MANUAL OF VENEREAL DISEASES. Being a Concise Description of those Affections and of their Treatment. Third Edition, post 8vo, 2s. 6d.

HINTS TO CANDIDATES FOR COMMISSIONS IN THE PUBLIC MEDICAL SERVICES, WITH EXAMINATION QUESTIONS, VOCABULARY OF HINDUSTANI MEDICAL TERMS, ETC. 8vo, 2s.

SIR W. JENNER, Bart., M.D.
Physician in Ordinary to H.M. the Queen, and to H.R.H. the Prince of Wales.

THE PRACTICAL MEDICINE OF TO-DAY: Two Addresses delivered before the British Medical Association, and the Epidemiological Society, (1869). Small 8vo, 1s. 6d.

C. M. JESSOP, M.R.C.P.
Associate of King's College, London; Brigade Surgeon H.M. British Forces.

ASIATIC CHOLERA, being a Report on an Outbreak of Epidemic Cholera in 1876 at a Camp near Murree in India. With map, demy 8vo, 2s. 6d.

GEORGE LINDSAY JOHNSON, M.A., M.B., B.C. CANTAB.
Clinical Assistant, late House Surgeon and Chloroformist, Royal Westminster Ophthalmic Hospital; Medical and Surgical Registrar, &c.

A **NEW METHOD OF TREATING CHRONIC GLAU-COMA**, based on Recent Researches into its Pathology. With Illustrations and coloured frontispiece, demy 8vo, 3s. 6d.

NORMAN W. KINGSLEY, M.D.S., D.D.S.
President of the Board of Censors of the State of New York; Member of the American Academy of Dental Science, &c.

A **TREATISE ON ORAL DEFORMITIES AS A BRANCH OF MECHANICAL SURGERY.** With over 350 Illustrations, 8vo, 16s.

E. A. KIRBY, M.D., M.R.C.S. ENG.
Late Physician to the City Dispensary.

I.

A **PHARMACOPŒIA OF SELECTED REMEDIES**, WITH THERAPEUTIC ANNOTATIONS, Notes on Alimentation in Disease, Air, Massage, Electricity and other Supplementary Remedial Agents, and a Clinical Index; arranged as a Handbook for Prescribers. Sixth Edition, enlarged and revised, demy 4to, 7s.

II.

ON THE VALUE OF PHOSPHORUS AS A REMEDY FOR LOSS OF NERVE POWER. Fifth Edition, 8vo, 2s. 6d.

J. WICKHAM LEGG, F.R.C.P.
Assistant Physician to Saint Bartholomew's Hospital, and Lecturer on Pathological Anatomy in the Medical School.

I.

ON THE BILE, JAUNDICE, AND BILIOUS DISEASES. With Illustrations in chromo-lithography, 719 pages, roy. 8vo, 25s.

II.

A **GUIDE TO THE EXAMINATION OF THE URINE**; intended chiefly for Clinical Clerks and Students. Fifth Edition, revised and enlarged, with additional Illustrations, fcap. 8vo, 2s. 6d.

III.

A **TREATISE ON HÆMOPHILIA, SOMETIMES CALLED THE HEREDITARY HÆMORRHAGIC DIATHESIS.** Fcap. 4to, 7s. 6d.

DR. GEORGE LEWIN.
Professor at the Fr. Wilh. University, and Surgeon-in-Chief of the Syphilitic Wards and Skin Disease Wards of the Charité Hospital, Berlin.

THE TREATMENT OF SYPHILIS WITH SUBCUTA-NEOUS SUBLIMATE INJECTIONS. Translated by DR. CARL PRŒGLE, and DR. E. H. GALE, late Surgeon United States Army. Small 8vo, 7s.

LEWIS'S PRACTICAL SERIES.

Under this title Mr. LEWIS purposes publishing a complete Series of Monographs, embracing the various branches of Medicine and Surgery.

The volumes, written by well-known Hospital Physicians and Surgeons recognized as authorities in the subjects of which they treat, are in active preparation. The works are intended to be of a THOROUGHLY PRACTICAL nature, calculated to meet the requirements of the general practitioner, and to present the most recent information in a compact and readable form; the volumes will be handsomely got up, and issued at low prices, varying with the size of the works.

BODILY DEFORMITIES AND THEIR TREATMENT: A HANDBOOK OF PRACTICAL ORTHOPÆDICS. By H. A. REEVES, F.R.C.S. Edin., Senior Assistant Surgeon and Teacher of Practical Surgery at the London Hospital; Surgeon to the Royal Orthopædic Hospital, &c. With numerous Illustrations, cr. 8vo. [In the press.

DENTAL SURGERY FOR GENERAL PRACTITIONERS AND STUDENTS OF MEDICINE. By ASHLEY W. BARRETT, M.B. Lond., M.R.C.S., L.D.S., Dental Surgeon to, and Lecturer on Dental Surgery and Pathology in the Medical School of, the London Hospital. With Illustrations, crown 8vo. [In the press.

*** Further volumes will be announced in due course.

LEWIS'S POCKET MEDICAL VOCABULARY.
[In the Press.

J. S. LOMBARD, M.D.
Formerly Assistant Professor of Physiology in Harvard College.

I.
EXPERIMENTAL RESEARCHES ON THE REGIONAL TEMPERATURE OF THE HEAD, under Conditions of Rest, Intellectual Activity and Emotion. With Illustrations, 8vo, 8s.

II.
ON THE NORMAL TEMPERATURE OF THE HEAD. 8vo, 5s.

WILLIAM THOMPSON LUSK, A.M., M.D.
Professor of Obstetrics and Diseases of Women in the Bellevue Hospital Medical College, &c.

THE SCIENCE AND ART OF MIDWIFERY. Second Edition, with numerous Illustrations, 8vo, 18s.

JOHN MACPHERSON, M.D.
Inspector-General of Hospitals H.M. Bengal Army (Retired). Author of "Cholera in its Home," &c.

ANNALS OF CHOLERA FROM THE EARLIEST PERIODS TO THE YEAR 1817. With a map. Demy 8vo, 7s. 6d.

DR. V. MAGNAN.
Physician to St. Anne Asylum, Paris; Laureate of the Institute.

ON ALCOHOLISM, the Various Forms of Alcoholic Delirium and their Treatment. Translated by W. S. GREENFIELD, M.D., M.R.C.P. 8vo, 7s. 6d.

A. COWLEY MALLEY, B.A., M.B., B.CH. T.C.D.

PHOTO-MICROGRAPHY; including a description of the Wet Collodion and Gelatino-Bromide Processes, together with the best methods of Mounting and Preparing Microscopic Objects for Photo-Micrography. Second Edition, with Photographs and Illustrations, crown 8vo. *[In the press.*

PATRICK MANSON, M.D., C.M.
Amoy, China.

THE FILARIA SANGUINIS HOMINIS; AND CERTAIN NEW FORMS OF PARASITIC DISEASE IN INDIA, CHINA, AND WARM COUNTRIES. Illustrated with Plates and Charts. 8vo, 10s. 6d.

PROFESSOR MARTIN.

MARTIN'S ATLAS OF OBSTETRICS AND GYNÆCOLOGY. Edited by A. Martin, Docent in the University of Berlin. Translated and edited with additions by Fancourt Barnes, M.D., M.R.C.P., Physician to the Chelsea Hospital for Women; Obstetric Physician to the Great Northern Hospital; and to the Royal Maternity Charity of London, &c. Medium 4to, Morocco half bound, 31s. 6d. net.

WILLIAM MARTINDALE, F.C.S.
Late Examiner of the Pharmaceutical Society, and late Teacher of Pharmacy and Demonstrator of Materia Medica at University College.

AND

W. WYNN WESTCOTT, M.B. LOND.
Deputy Coroner for Central Middlesex.

THE EXTRA PHARMACOPŒIA of Unofficial Drugs and Chemical and Pharmaceutical Preparations, with References to their Use abstracted from the Medical Journals and a Therapeutic Index of Diseases and Symptoms. Third Edition, revised with numerous additions, limp roan, med. 24mo, 7s., and an edition in fcap. 8vo, with room for marginal notes, cloth, 7s. *[Now ready.*

J. F. MEIGS, M.D.
Consulting Physician to the Children's Hospital, Philadelphia.

AND

W. PEPPER, M.D.
Lecturer on Clinical Medicine in the University of Pennsylvania.

A PRACTICAL TREATISE ON THE DISEASES OF CHILDREN. Seventh Edition, revised and enlarged, roy. 8vo, 28s.

Wm. JULIUS MICKLE, M.D., M.R.C.P. LOND.
Member of the Medico-Psychological Association of Great Britain and Ireland; Member of the Clinical Society, London; Medical Superintendent, Grove Hall Asylum, London.

GENERAL PARALYSIS OF THE INSANE. 8vo, 10s.

KENNETH W. MILLICAN, B.A. CANTAB., M.R.C.S.

THE EVOLUTION OF MORBID GERMS: A Contribution to Transcendental Pathology. Cr. 8vo, 3s. 6d.

E. A. MORSHEAD, M.R.C.S., L.R.C.P.
Assistant to the Professor of Medicine in University College, London.

TABLES OF THE PHYSIOLOGICAL ACTION OF DRUGS. Fcap. 8vo, 1s.

A. STANFORD MORTON, M.B., F.R.C.S. ED.
Senior Assistant Surgeon, Royal South London Ophthalmic Hospital.

REFRACTION OF THE EYE: Its Diagnosis, and the Correction of its Errors, with Chapter on Keratoscopy. Second edit., with Illustrations, small 8vo, 2s. 6d.

WILLIAM MURRELL, M.D., F.R.C.P.
Lecturer on Materia Medica and Therapeutics at Westminster Hospital; Examiner in Materia Medica, University of Edinburgh.

I.
WHAT TO DO IN CASES OF POISONING. Fourth Edition, revised and enlarged, royal 32mo, 3s. 6d.

II.
NITRO-GLYCERINE AS A REMEDY FOR ANGINA PECTORIS. Crown 8vo, 3s. 6d.

WILLIAM NEWMAN, M.D. LOND., F.R.C.S.
Surgeon to the Stamford Infirmary.

SURGICAL CASES: Mainly from the Wards of the Stamford, Rutland, and General Infirmary, 8vo, paper boards, 4s. 6d.

DR. FELIX von NIEMEYER.
Late Professor of Pathology and Therapeutics; Director of the Medical Clinic of the University of Tübingen.

A TEXT-BOOK OF PRACTICAL MEDICINE, WITH PARTICULAR REFERENCE TO PHYSIOLOGY AND PATHOLOGICAL ANATOMY. Translated from the Eighth German Edition, by special permission of the Author, by GEORGE H. HUMPHREY, M.D., and CHARLES E. HACKLEY, M.D., Revised Edition, 2 vols., large 8vo, 36s.

C. F. OLDHAM, M.R.C.S., L.R.C.P.
Surgeon H.M. Indian Forces; late in Medical charge of the Dalhousie Sanitarium.

WHAT IS MALARIA? and why is it most intense in hot climates? An explanation of the Nature and Cause of the so-called Marsh Poison, with the Principles to be observed for the Preservation of Health in Tropical Climates and Malarious Districts. Demy 8vo, 7s. 6d.

G. OLIVER, M.D., M.R.C.P.

I.
THE HARROGATE WATERS: Data Chemical and Therapeutical, with notes on the Climate of Harrogate. Addressed to the Medical Profession. Crown 8vo, with Map of the Wells, 3s. 6d.

II.
ON BEDSIDE URINE TESTING: including Quantitative Albumen and Sugar. Third edition, revised and enlarged, fcap. 8vo.
[*In the press.*

JOHN S. PARRY, M.D.
Obstetrician to the Philadelphia Hospital, Vice-President of the Obstetrical and Pathological Societies of Philadelphia, &c.

EXTRA-UTERINE PREGNANCY; Its Causes, Species, Pathological Anatomy, Clinical History, Diagnosis, Prognosis and Treatment. 8vo, 8s.

E. RANDOLPH PEASLEE, M.D., LL.D.
Late Professor of Gynæcology in the Medical Department of Dartmouth College; President of the New York Academy of Medicine, &c., &c.

OVARIAN TUMOURS: Their Pathology, Diagnosis, and Treatment, especially by Ovariotomy. Illustrations, roy. 8vo, 16s.

G. V. POORE, M.D., F.R.C.P.
Professor of Medical Jurisprudence, University College; Assistant Physician to, and Physician in charge of the Throat Department of University College Hospital.

LECTURES ON THE PHYSICAL EXAMINATION OF THE MOUTH AND THROAT. With an Appendix of Cases. 8vo, 3s. 6d.

R. DOUGLAS POWELL, M.D., F.R.C.P. LOND.
Physician to the Middlesex Hospital, and Physician to the Hospital for Consumption and Diseases of the Chest at Brompton.

DISEASES OF THE LUNGS AND PLEURÆ. Third Edition, rewritten and enlarged. With Illustrations, 8vo.
[*In preparation.*

AMBROSE L. RANNEY, A.M., M.D.
Adjunct Professor of Anatomy in the University of New York, etc.

THE APPLIED ANATOMY OF THE NERVOUS SYS-
TEM, being a study of this portion of the Human Body from a standpoint of its general interest and practical utility, designed for use as a Text-book and a Work of Reference. With 179 Illustrations, 8vo, 20s.

H. A. REEVES, F.R.C.S. ED.
Senior Assistant Surgeon and Teacher of Practical Surgery at the London Hospital; Surgeon to the Royal Orthopædic Hospital, &c.

BODILY DEFORMITIES AND THEIR TREATMENT:
A Handbook of Practical Orthopædics. With numerous Illustrations, crown 8vo. [*In the press.*
Lewis's Practical Series].

RALPH RICHARDSON, M.A., M.D.
Fellow of the College of Physicians, Edinburgh.

ON THE NATURE OF LIFE: An Introductory Chap-
ter to Pathology. Second Edition, revised and enlarged. Fcap. 4to, 10s. 6d.

W. RICHARDSON, M.A., M.D., M.R.C.P.

REMARKS ON DIABETES, ESPECIALLY IN REFER-
ENCE TO TREATMENT. Demy 8vo, 4s. 6d.

SYDNEY RINGER, M.D.
Professor of the Principles and Practice of Medicine in University College; Physician to, and Professor of Clinical Medicine in, University College Hospital.

I.
A HANDBOOK OF THERAPEUTICS. Tenth Edition, 8vo, 15s.

II.
ON THE TEMPERATURE OF THE BODY AS
A MEANS OF DIAGNOSIS AND PROGNOSIS IN PHTHISIS. Second Edition, small 8vo, 2s. 6d.

FREDERICK T. ROBERTS, M.D., B.SC., F.R.C.P.
Examiner in Medicine at the Royal College of Surgeons; Professor of Therapeutics in University College; Physician to University College Hospital; Physician to Brompton Consumption Hospital, &c.

I.
A HANDBOOK OF THE THEORY AND PRACTICE
OF MEDICINE. Fifth Edition, with Illustrations, in one volume, large 8vo, 21s.

II.
NOTES ON MATERIA MEDICA AND PHARMACY. Fcap. 8vo, 7s. 6d. [*Now ready.*

D. B. St. JOHN ROOSA, M.A., M.D.
Professor of Diseases of the Eye and Ear in the University of the City of New York; Surgeon to the Manhattan Eye and Ear Hospital; Consulting Surgeon to the Brooklyn Eye and Ear Hospital, &c., &c.

A PRACTICAL TREATISE ON THE DISEASES OF THE EAR, including the Anatomy of the Organ. Fourth Edition, Illustrated by wood engravings and chromo-lithographs, large 8vo, 22s.

J. BURDON SANDERSON, M.D., LL.D., F.R.S.
Jodrell Professor of Physiology in University College, London.

UNIVERSITY COLLEGE COURSE OF PRACTICAL EXERCISES IN PHYSIOLOGY. With the co-operation of F. J. M. PAGE, B.Sc., F.C.S.; W. NORTH, B.A., F.C.S., and AUG. WALLER, M.D. Demy 8vo, 3s. 6d.

W. H. O. SANKEY, M.D. LOND., F.R.C.P.
Late Lecturer on Mental Diseases, University College and School of Medicine for Women, London; Formerly Medical Superintendent (Female Department) of Hanwell Asylum; President of Medico-Psychological Society, &c.

LECTURES ON MENTAL DISEASE. Second Edition, with coloured plates, 8vo, 12s. 6d. [*Now ready.*

ALDER SMITH, M.B. LOND., F.R.C.S.
Resident Medical Officer, Christ's Hospital, London.

RINGWORM: Its Diagnosis and Treatment. Second Edition, rewritten and enlarged. With Illustrations, fcap. 8vo, 4s. 6d.

J. LEWIS SMITH, M.D.
Physician to the New York Infants' Hospital; Clinical Lecturer on Diseases of Children in Bellevue Hospital Medical College.

A TREATISE ON THE DISEASES OF INFANCY AND CHILDHOOD. Fifth Edition, with Illustrations, large 8vo, 21s.

FRANCIS W. SMITH, M.B., B.S.

THE LEAMINGTON WATERS; CHEMICALLY, THERAPEUTICALLY AND CLINICALLY CONSIDERED; with observations on the climate of Leamington. With Illustrations, crown 8vo, 2s. 6d.

JAMES STARTIN, M.B., M.R.C.S.
Surgeon and Joint Lecturer to St. John's Hospital for Diseases of the Skin.

LECTURES ON THE PARASITIC DISEASES OF THE SKIN. VEGETOID AND ANIMAL. With Illustrations, Crown 8vo, 3s. 6d.

HENRY R. SWANZY, M.A., M.B., F.R.C.S.I.
Examiner in Ophthalmic Surgery, University of Dublin; Surgeon to the National Eye and Ear Infirmary, Dublin; Ophthalmic Surgeon at the Adelaide Hospital, Dublin.

HANDBOOK OF DISEASES OF THE EYE AND THEIR TREATMENT. Illustrated with wood-engravings, colour tests, etc., large post 8vo, 10s. 6d. [*Now ready.*

LEWIS A. STIMSON, B.A., M.D.
Surgeon to the Presbyterian Hospital; Professor of Pathological Anatomy in the Medical Faculty of the University of the City of New York.

A MANUAL OF OPERATIVE SURGERY. With three hundred and thirty-two Illustrations. Post 8vo, 10s. 6d.

HUGH OWEN THOMAS, M.R.C.S.

I.

DISEASES OF THE HIP, KNEE, AND ANKLE JOINTS, with their Deformities, treated by a new and efficient method. With an Introduction by RUSHTON PARKER, F.R.C.S, Lecturer on Surgery at the School of Medicine, Liverpool. Second Edition, 8vo, 25s.

II.

CONTRIBUTIONS TO MEDICINE AND SURGERY:—

PART 1.—Intestinal Obstruction; with an Appendix on the Action of Remedies. 10s.
  „ 2.—The Principles of the Treatment of Joint Disease, Inflammation, Anchylosis, Reduction of Joint Deformity, Bone Setting. 5s.
  „ 5.—On Fractures of the Lower Jaw. 1s.
  „ 8.—The Inhibition of Nerves by Drugs. Proof that Inhibitory Nerve-Fibres do not exist. 1s.

(Parts 3, 4, 6, 7, 9, 10, are expected shortly).

J. ASHBURTON THOMPSON, M.R.C.S.
Late Surgeon at King's Cross to the Great Northern Railway Company

FREE PHOSPHORUS IN MEDICINE WITH SPECIAL REFERENCE TO ITS USE IN NEURALGIA. A contribution to Materia Medica and Therapeutics. An account of the History, Pharmaceutical Preparations, Dose, Internal Administration, and Therapeutic uses of Phosphorus; with a Complete Bibliography of this subject, referring to nearly 200 works upon it. Demy 8vo, 7s. 6d.

J. C. THOROWGOOD, M.D.
Assistant Physician to the City of London Hospital for Diseases of the Chest.

THE CLIMATIC TREATMENT OF CONSUMPTION AND CHRONIC LUNG DISEASES. Third Edition, post 8vo, 3s 6d.

EDWARD T. TIBBITS, M.D. LOND.
Physician to the Bradford Infirmary; and to the Bradford Fever Hospital.

MEDICAL FASHIONS IN THE NINETEENTH CENTURY, including a Sketch of Bacterio-Mania and the Battle of the Bacilli. Crown 8vo, 2s. 6d.

LAURENCE TURNBULL, M.D., PH.G.
Aural Surgeon to Jefferson Medical College Hospital, &c., &c.

ARTIFICIAL ANÆSTHESIA: A Manual of Anæsthetic Agents, and their Employment in the Treatment of Disease. Second Edition, with Illustrations, crown 8vo, 6s.

W. H. VAN BUREN, M.D., LL.D.
Professor of Surgery in the Bellevue Hospital Medical College.

DISEASES OF THE RECTUM: And the Surgery of the Lower Bowel. Second Edition, with Illustrations, 8vo, 14s.

RUDOLPH VIRCHOW, M.D.
Professor in the University, and Member of the Academy of Sciences of Berlin, &c., &c.

INFECTION-DISEASES IN THE ARMY, Chiefly Wound Fever, Typhoid, Dysentery, and Diphtheria. Translated from the German by JOHN JAMES, M.B., F.R.C.S. Fcap. 8vo, 1s. 6d.

ALFRED VOGEL, M.D.
Professor of Clinical Medicine in the University of Dorpat, Russia.

A PRACTICAL TREATISE ON THE DISEASES OF CHILDREN. Translated and Edited by H. RAPHAEL, M.D. From the Fourth German Edition, illustrated by six lithographic plates, part coloured, large 8vo, 18s.

A. DUNBAR WALKER, M.D., C.M.

THE PARENT'S MEDICAL NOTE BOOK. Oblong post 8vo, cloth, 1s. 6d.

W. SPENCER WATSON, F.R.C.S. ENG., B.M. LOND.
Surgeon to the Great Northern Hospital; Surgeon to the Royal South London Ophthalmic Hospital.

I.
DISEASES OF THE NOSE AND ITS ACCESSORY CAVITIES. Profusely Illustrated. Demy 8vo, 18s.

II.
EYEBALL-TENSION: Its Effects on the Sight and its Treatment. With woodcuts, p. 8vo, 2s. 6d.

III.
ON ABSCESS AND TUMOURS OF THE ORBIT. Post 8vo, 2s. 6d.

A. DE WATTEVILLE, M.A., M.D., B.SC., M.R.C.S.
Physician in Charge of the Electro-therapeutical Department at St. Mary's Hospital.

A PRACTICAL INTRODUCTION TO MEDICAL ELECTRICITY. Second Edition, re-written and enlarged, copiously Illustrated, 8vo, 9s. [*Just published.*

FRANCIS H. WELCH, F.R.C.S.
Surgeon Major, A.M.D.

ENTERIC FEVER: as Illustrated by Army Data at Home and Abroad, its Prevalence and Modifications, Ætiology, Pathology and Treatment. 8vo, 5s. 6d. [*Just published.*

DR. F. WINCKEL.
Formerly Professor and Director of the Gynæcological Clinic at the University of Rostock.

THE PATHOLOGY AND TREATMENT OF CHILD-BED: A Treatise for Physicians and Students. Translated from the Second German edition, with many additional notes by the Author, by J. R. CHADWICK, M.D., 8vo, 14s.

EDWARD WOAKES, M.D. LOND.

Senior Aural Surgeon and Lecturer on Aural Surgery at the London Hospital; Senior Surgeon to the Hospital for Diseases of the Throat.

ON DEAFNESS, GIDDINESS AND NOISES IN THE HEAD.

VOL. I.—POST-NASAL CATARRH, AND DISEASES OF THE NOSE CAUSING DEAFNESS. With Illustrations, cr. 8vo, 6s. 6d.

VOL. II.—ON DEAFNESS, GIDDINESS AND NOISES IN THE HEAD. Third Edition, with Illustrations, cr. 8vo. [*In preparation.*

E. T. WILSON, B.M. OXON., F.R.C.P. LOND.
Physician to the Cheltenham General Hospital and Dispensary.

DISINFECTANTS AND HOW TO USE THEM. In Packets of one doz. price 1s.

Clinical Charts For Temperature Observations, etc.
Arranged by W. RIGDEN, M.R.C.S. Price 7s. per 100, or 1s. per dozen.

Each Chart is arranged for four weeks, and is ruled at the back for making notes of cases; they are convenient in size, and are suitable both for hospital and private practice.

PERIODICAL WORKS PUBLISHED BY H. K. LEWIS.

THE NEW SYDENHAM SOCIETY'S PUBLICATIONS. Annual Subscription, One Guinea.
(Report of the Society, with Complete List of Works and other information, gratis on application.)

ARCHIVES OF PEDIATRICS. A Monthly Journal, devoted to the Diseases of Infants and Children. Annual Subscription, 12s. 6d., post free.

THE NEW YORK MEDICAL JOURNAL. A Weekly Review of Medicine. Annual Subscription, One Guinea, post free.

THE THERAPEUTIC GAZETTE. A Monthly Journal, devoted to the Science of Pharmacology, and to the introduction of New Therapeutic Agents. Annual Subscription, 5s., post free.

THE GLASGOW MEDICAL JOURNAL. Published Monthly. Annual Subscription, 20s., post free. Single numbers, 2s. each.

LIVERPOOL MEDICO-CHIRURGICAL JOURNAL, including the Proceedings of the Liverpool Medical Institution. Published twice yearly, 3s. 6d. each.

THE INDIAN MEDICAL JOURNAL. A Journal of Medical and Sanitary Science specially devoted to the Interests of the Medical Services. Annual Subscription, 24s., post free.

THE MIDLAND MEDICAL MISCELLANY AND PROVINCIAL MEDICAL JOURNAL. Annual Subscription, 7s. 6d., post free.

TRANSACTIONS OF THE COLLEGE OF PHYSICIANS OF PHILADELPHIA. Volumes I to VI., now ready, 8vo, 10s. 6d. each.

*** MR. LEWIS has transactions with the leading publishing firms in America for the sale of his publications in that country. Arrangements are made in the interests of Authors either for sending a number of copies of their works to the United States, or having them reprinted there, as may be most advantageous.

Mr. Lewis's publications can be procured of any bookseller in any part of the world.

www.ingramcontent.com/pod-product-compliance
Lightning Source LLC
Chambersburg PA
CBHW020757230426
43666CB00007B/738